Low Power Networks-on-Chip

Cristina Silvano • Marcello Lajolo
Gianluca Palermo
Editors

Low Power Networks-on-Chip

 Springer

Editors

Cristina Silvano
Politecnico di Milano
Dip. Elettronica e Informazione (DEI)
Via Ponzio 34/5
20133 Milano
Italy
silvano@elet.polimi.it

Gianluca Palermo
Politecnico di Milano
Dip. Elettronica e Informazione (DEI)
Via Ponzio 34/5
20133 Milano
Italy
gpalermo@elet.polimi.it

Marcello Lajolo
NEC Laboratories America, Inc.
Independence Way 4
08540 Princeton New Jersey
USA
lajolo@nec-labs.com

ISBN 978-1-4899-9437-0   ISBN 978-1-4419-6911-8 (eBook)
DOI 10.1007/978-1-4419-6911-8
Springer New York Dordrecht Heidelberg London

Printed on acid-free paper

Springer is part of Springer Science+Business Media (www.springer.com)

"Ma sopra tutte le invenzioni stupende, qual eminenza fu quella di colui che s'immaginò di trovar modo di comunicare i suoi più reconditi pensieri a qualsivoglia altra persona, benché distante per lunghissimo intervallo di luogo e di tempo? Parlare con quelli che son nell'Indie, parlare a quelli che non sono ancora nati né saranno se non di qua a mille e diecimila anni? E con qual facilità? Con i vari accozzamenti di venti caratteruzzi sopra una carta."

Galileo Galilei, *Dialogo sopra i due massimi sistemi del mondo, Tolemaico e Copernicano, Firenze, 1632*

"But surpassing all stupendous inventions, what sublimity of mind was his who dreamed of finding means to communicate his deepest thoughts to any other person, though distant by mighty intervals of place and time! Of talking with those who are in India; of speaking to those who are not yet born and will not be born for a thousand or ten thousand years; and with what facility, by the different arrangements of twenty characters upon a page."

Galileo Galilei, *Dialogue concerning the two chief world systems, Ptolemaic & Copernican, Florence, 1632*

Preface

Given the increasing complexity of multiprocessor system-on-chip (MPSoC) designs, the current trends in on-chip communication architectures are converging towards the network-on-chip (NoC). The NoC-based design approach represents a high bandwidth and low energy solution. Using the NoC-based design approach has several other advantages, such as scalability, reliability, IP reusability and separation of IP design from on-chip communication design and interfacing. NoC design represents a new paradigm to design MPSoC, shifting the design methodologies from computation-based to communication-based.

Given these premises, during the last decade, we assisted an increasing research effort on NoC architectures and related design methodologies. Many key design challenges of NoC have been investigated in the past years. These challenges have recently been classified by Marculescu et al. in three main categories: the design of the communication infrastructure, the selection of the communication paradigm and the application mapping optimization. First, the problem of designing the communication infrastructure consists in turn of the following problems: network topology synthesis, the selection of the channel width, the buffer sizing problem and the NoC floorplanning problem. Second, the selection of the communication paradigm includes the routing problem and the choice of switching techniques (store-and-forward, cut-through, wormhole, etc.) to be used. Third, the application mapping optimization problem consists in turn of the IP mapping and the task scheduling problems of an application onto the NoC platform. All these optimisation techniques should take into consideration several metrics of interest to be traded off. These metrics are mainly performance, energy, quality of service, reliability and security.

In this scenario, even some semiconductor industries have started to propose some NoC-based designs. Among them, we can cite the Aetheral NoC from NXP-Philips, the STNoC from STMicroelectronics and the 80-core NoC from Intel. Several industrial design flows supporting NoC design have also been proposed, such as the CHAIN works tool suite by Silistix, the NoCexplorer and NoCcompiler frameworks by Arteris and the iNOCs tools from iNoCs. The interest demonstrated by several industries and EDA providers contributed to confirm NoC as a feasible and energy-efficient approach to interconnect a scalable number of IP cores on a single die.

Although many scientific books and journal papers have recently been published, many challenging topics related to NoC research are still open. The story behind this book began more than a year ago, when we started thinking with Charles Glaser from Springer about a book focusing on low-power NoC, as power and energy issues still represent one of the limiting factors in integrating multi- and many-cores on a single chip. Power-aware design techniques at several abstraction levels represent the enabling keys for an energy-efficient design of on-chip interconnection network. Starting from this idea, the book tries to answer to the necessity of a single textbook on the topic of low-power NoC, covering power- and energy-aware design techniques from several perspectives and abstraction levels. To this purpose, the present book tries to put together several outstanding contributions in several areas of low-power NoC design.

The book chapters are organized in three parts. In Part I, several power-aware design techniques on NoC are discussed from the low-level perspective. These low-level NoC design techniques address the following topics: hybrid circuit/packet switched networks, run-time power-gating techniques, adaptive voltage control techniques for NoC links and asynchronous communication. In Part II, several system-level power-aware design techniques are presented dealing with application-specific routing algorithms, adaptive data compression and design techniques for latency constrained and power optimized NoCs. In Part III of the book, some emerging technologies related to low-power NoC, namely 3D stacking, CMOS nanophotonics and RF-interconnect are discussed to envision their applicability to meet the requirements imposed by future NoC architectures.

Entering Part I on low-level design techniques, Chap. 1 introduces some issues and challenges for future NoCs with demands for high bandwidth and low energy. Starting from the analysis of some state-of-the-art approaches to design NoC architectures, the chapter presents details of how coupling packet-switched arbitration with circuit-switched data transfer can achieve energy savings and improve network efficiency by reducing arbitration overhead and increasing overall utilisation of network resources. In this hybrid network, packet-switched arbitration is used to reserve future circuit-switched channels for the data transfer, thus eliminating the performance bottlenecks associated with pure circuit-switched networks, while maintaining their power advantage. Furthermore, the chapter discusses how proximity-based data streaming can increase network throughput and improve energy efficiency. Finally, some NoC measurements and design trade offs are analysed on 45 nm CMOS technology from an industrial research perspective.

Chapter 2 surveys several power gating techniques to reduce the leakage power of on-chip routers. Leakage power is responsible for a considerable portion of the active power in recent process technologies. Then, the chapter introduces a run-time fine-grained power gating router, in which power supply to each router component (e.g. virtual-channel buffer, crossbar's multiplexer, and output latch) can be individually controlled in response to the applied workload. To mitigate the impact of wake-up latency of each power domain on application performance, the chapter introduces and discusses three wake-up control methods. Finally, the fine-grained power gating router with 35 micro power domains and the early wake-up methods are designed with a commercial 65 nm process and evaluated in terms of the area overhead, application performance and leakage power reduction.

Chapter 3 surveys the state of the art in energy-efficient communication link design for NoCs. After reviewing techniques at the datalink and physical abstraction layers, the chapter introduces a lookahead-based transition-aware adaptive voltage control method for achieving improved energy efficiency at moderate cost in performance and reliability. Then, performance and limitations of the proposed method are evaluated and future prospects in energy-efficient link design are projected.

Chapter 4 provides an overview of the various asynchronous techniques that are used at the link layer in NoCs, including signalling schemes, data encoding and synchronization solutions. These asynchronous techniques are discussed with a view of comparison in terms of area, power and performance. The fundamental issues of the formation of data tokens based on the principles of data validity, acknowledgement, delay-insensitivity, timing assumptions and soft-error tolerance are considered. The chapter also covers some of the aspects related to combining asynchronous communication links to form parts of the entire network architecture, which involves asynchronous logic for arbitration and routing hardware. To this end, the chapter also presents basic techniques for building small-scale controllers using the formal models of Petri nets and signal transition Graphs.

Entering Part II on system-level design techniques, Chap. 5 describes how the routing algorithm can be optimized in NoC platforms. Routing algorithm has a major effect on the performance (packet latency and throughput) as well as power consumption of NoC. A methodology to develop efficient and deadlock free routing algorithms which are specialized for an application or a set of concurrent applications is presented. The methodology, called application specific routing algorithms (APSRA), exploits the application-specific information regarding pairs of cores which communicate and other pairs which never communicate in the NoC platform. This information is used to maximize the adaptivity of the routing algorithm without compromising the important property of deadlock freedom. The chapter also presents an extensive comparison between the routing algorithms generated using APSRA methodology with general purpose deadlock-free routing algorithms. The simulation-based evaluations are performed using both synthetic traffic as well as traffic from real applications. The comparison embraces several performance indices such as degree of adaptiveness, average delay, throughput, power dissipation and energy consumption. In spite of an adverse impact on router architecture, the chapter proves that the higher adaptivity of APSRA leads to significant improvements in both routing performance and energy consumption.

Chapter 6 presents a method to exploit a table-based data compression technique, relying on value patterns in cache traffic. Compressing a large packet into a small one saves power consumption by reducing required operations in network components and decreases contention by increasing the effective bandwidth of shared resources. The main challenges are providing a scalable implementation of tables and minimizing the latency overhead of compression. We propose a shared table scheme that needs one encoding and one decoding table for each processing element, and a management protocol that does not require in-order delivery. This scheme eliminates table size dependence on a network size, which realizes scalability and reduces overhead cost of table for compression. The chapter also presents some simulation results obtained by using the proposed compression method for 8-core and 16-core tiled designs. The experimental results are discussed in terms of packet latency and network power consumption.

Chapter 7 describes the design process of a Network on-Chip for a high-end commercial system on-chip (SoC) application. Several design choices are discussed in the chapter focusing on the power optimization of the NoC while achieving the required performance. The chapter describes the NoC design steps including module mapping and allocation of customized capacities to links. Unlike previous studies, in which point-to-point, per-flow timing constraints were used, the chapter demonstrates the importance of using the application end-to-end traversal latency requirements during the optimization process. To compare several design alternatives, the chapter reports the synthesis results of an NoC design that meets the actual throughput and timing requirements of a commercial 4G SoC.

Entering Part III on future and emerging technologies, Chap. 8 addresses the problem of 2D and 3D SoC designs where the cores are grouped into voltage islands. To reduce the leakage power consumption, an island containing cores that are not used in an application can be shutdown, while the other islands can still be operational. When one or more of the islands are shutdown, the interconnect should allow the communication between islands that are operational. For this, the NoC has to be designed efficiently to allow shutdown of voltage islands, thereby reducing the leakage power consumption. The chapter presents methods to design NoC topologies that provide such a support for both 2D and 3D technologies. The chapter outlines how the concept of voltage islands needs to be considered during the topology synthesis phase itself. The chapter also analyses the benefits of migrating to 3D stacked chips for realistic applications that have multiple voltage islands.

Chapter 9 introduces the emerging CMOS nanophotonic technologies representing a compelling alternative to traditional all-electronic NoCs. This is because of nanophotonic NoCs can provide both higher throughput and lower power consumption than all-electrical NoCs. The chapter introduces CMOS nanophotonic technology and considers its use in photonic chip-wide networks enabling many-core microprocessors with greatly enhanced performance and flexibility while consuming less power than their electrical counterparts. The chapter also provides, as a case study, a design that takes advantage of CMOS nanophotonics to achieve ten-teraop performance in a 256-core 3D chip stack, using optically connected main memory, very high memory bandwidth, cache coherence across all cores, no bisection bandwidth limits on communication and cross-chip communication at very low latency with cache-line granularity.

Chapter 10 explores the use of multi-band RF-interconnect for future Network-on-Chip. RF-interconnect can communicate simultaneously through multiple frequency bands with low-power signal transmission and reconfigurable bandwidth. At the same time, the chapter investigates the CMOS mixed-signal circuit implementation challenges for improving the RF-I signalling integrity and efficiency.

Furthermore, the chapter proposes a micro-architectural framework that can be used to facilitate the exploration of scalable low-power NoC architectures based on physical planning and prototyping.

Due to the large number of topics discussed in the book and their heterogeneity, the background on low-power NoC is discussed chapter by chapter with a separate reference set for each chapter. This choice also contributed to make each chapter self-contained.

Overall, we believe that the chapters cover a set of definitely important and timely issues impacting the present and future research on low-power NoC. We sincerely hope that the book could become a solid reference in the next years. In our vision, the authors put a big effort in clearly presenting their technical contribution outlining the potential impact and some case studies. We would like to have the opportunity to specially thank all the authors who contributed to the book. A special thanks to Charles Glaser from Springer for encouraging us from the beginning of this book and Amanda Davis from Springer for her continuous support in reviewing the materials.

Milano, Italy	*Cristina Silvano*
Princeton, NJ, USA	*Marcello Lajolo*
Milano, Italy	*Gianluca Palermo*
April 2010	The Editors

About the Editors

Cristina Silvano received the M.S. degree in electronic engineering from Politecnico di Milano, Milano, Italy, in 1987 and the Ph.D. degree in computer engineering from the University of Brescia, Brescia, Italy, in 1999. From 1987 to 1996, she was a Senior Design Engineer with the R&D Labs, Groupe Bull, Pregnana, Italy. From 2000 to 2002, she was an Assistant Professor with the Department of Computer Science, University of Milan, Milano. She is currently an Associate Professor (with tenure) in computer engineering with the Dipartimento di Elettronica e Informazione, Politecnico di Milano. She has published one scientific international book and more than 70 papers in international journals and conference proceedings, and she is the holder of several international patents. Her primary research interests are in the area of computer architectures and computer-aided design of digital systems, with particular emphasis on design space exploration and low-power design techniques for multiprocessor systems-on-chip. She participated in several national and international research projects, some of them in collaboration with STMicroelectronics. She is currently the European Coordinator of the project FP7-2PARMA-248716 on "PARallel PAradigms and Run-time MAnagement techniques for Many-core Architectures" (Jan 2010 to Dec 2012). She is also the European Coordinator of the on-going project FP7-MULTICUBE-216693 on "Multi-objective design space exploration of multiprocessor SoC architectures for embedded multimedia applications" (Jan 2008 to June 2010).

Marcello Lajolo received his Master and Ph.D. degrees in electrical engineering, both from Politecnico di Torino (Italy) in 1995 and 1999, respectively. He then joined the Computer & Communication Research Laboratories (CCRL; now NEC Laboratories America) in Princeton, NJ, where he led various projects in the areas of on-chip communication design and advanced embedded architectures. He also collaborates with Advanced Learning and Research Institute (ALaRI) in Lugano, Switzerland, where he has been teaching a course on networks on chip since 2002. He has served or is serving as a program committee member for major conferences in electronic design automation and embedded system design like DAC, DATE, ASP-DAC, and ISCAS. He has given full-day tutorials at conferences like ICCAD, ASP-DAC, ICCD and others in the area of embedded system design. He is a Senior

Member of the IEEE. His primary research topics are related to Networks on Chip, Hardware/Software Codesign Low Power Design, Computer Architectures, High Level Synthesis of Digital Integrated Circuits and System-on-Chip Testing.

Gianluca Palermo received the M.S. degree in electronic engineering and the Ph.D. degree in computer engineering from Politecnico di Milano, Milano, Italy, in 2002 and 2006, respectively. Previously, he was a Consultant Engineer with the Low Power Design Group, Advanced System Technology, STMicroelectronics, where he worked on network-on-chip, and also a Research Assistant with the Advanced Learning and Research Institute, University of Lugano, Lugano, Switzerland. He is currently an Assistant Professor in the Dipartimento di Elettronica e Informazione, Politecnico di Milano. His research interests include design methodologies and architectures for embedded systems, focusing on low-power design, on-chip multiprocessors, and network-on-chip. He participated in several national and international research projects.

Contents

Part III   Future and Emerging Technologies

Contributors

Jung Ho Ahn Seoul National University, Sumon, Gyeonggi-do, Korea, gajh@snu.ac.kr

Hideharu Amano Keio University, 3-14-1, Hiyoshi, Kohoku-ku, Yokohama, Japan 223-8522, hunga@am.ics.keio.ac.jp

Paul Ampadu University of Rochester, Rochester, NY 14627, USA, ampadu@ece.rochester.edu

Mark A. Anders Intel Corporation, Hillsboro, OR, USA, mark.a.anders@intel.com

Raymond G. Beausoleil Hewlett-Packard Labs, Palo Alto, CA, USA, ray.beausoleil@hp.com

Luca Benini DEIS, Univerity of Bologna, Bologna, Italy, lbenini@deis.unibo.it

Rudy Beraha Qualcomm Corp. Research and Development, San Diego, California 92121, USA, rberaha@qualcomm.com

Nathan Binkert Hewlett-Packard Labs, Palo Alto, CA, USA, binkert@hp.com

Shekhar Y. Borkar Intel Corporation, Hillsboro, OR, USA, shekhar.y.borkar@intel.com

Vincenzo Catania Dipartimento di Ingegneria Informatica e delle Telecomunicazioni, University of Catania, Italy, vcatania@diit.unict.it

Mau-Chung Frank Chang Electrical Engineering Department, University of California, Los Angeles, Engineering IV Building, CA 90095, Los Angeles, USA, mfchang@ee.ucla.edu

Israel Cidon Electrical Engineering Department, Technion - Israel Institute of Technology, Haifa 32000, Israel, cidon@ee.technion.ac.il

Jason Cong Computer Science Department, University of California, Los Angeles, 4731J, Boelter Hall, Los Angeles, CA 90095, cong@cs.ucla.edu

Giovanni De Micheli LSI, EPFL, Lausanne, Switzerland, giovanni.demicheli@epfl.ch

Al Davis Hewlett-Packard Labs, Palo Alto, CA, USA, ald@hp.com

Marco Fiorentino Hewlett-Packard Labs, Palo Alto, CA, USA, marco.fiorentino@hp.com

Bo Fu University of Rochester, Rochester, NY 14627, USA, bofu@ece.rochester.edu

Stanislavs Golubcovs Asynchronous Systems Laboratory, School of EECE, Newcastle University, Newcastle upon Tyne, United Kingdom, stanislavs.golubcovs@ncl.ac.uk

Rickard Holsmark Department of Electronics and Computer Engineering, Jönköping University, Jönköping, Sweden, Rickard.Holsmark@jth.hj.se

Yuho Jin Department of Electrical Engineering, University of Southern California, 3740 McClintock Ave., Los Angeles, CA 90089, USA, yujin@usc.edu

Norman P. Jouppi Hewlett-Packard Labs, Palo Alto, CA, USA, norm.jouppi@hp.com

Himanshu Kaul Intel Corporation, Hillsboro, OR, USA, himanshu.kaul@intel.com

Eun Jung Kim Department of Computer Science and Engineering, Texas A&M University, College Station, TX 77843-3112, USA ejkim@cse.tamu.edu

Michihiro Koibuchi National Institute of Informatics, 2-1-2, Hitotsubashi, Chiyoda-ku, Tokyo, Japan 101-8430, koibuchi@nii.ac.jp

Avinoam Kolodny Electrical Engineering Department, Technion - Israel Institute of Technology, Haifa 32000, Israel, kolodny@ee.technion.ac.il

Ram K. Krishnamurthy Intel Corporation, Hillsboro, OR, USA, ram.krishnamurthy@intel.com

Shashi Kumar Department of Electronics and Computer Engineering, Jönköping University, Jönköping, Sweden, Shashi.Kumar@jth.hj.se

Hiroki Matsutani The University of Tokyo, 7-3-1 Hongo, Bunkyo-ku, Tokyo, Japan 113-8656, matsutani@hal.ipc.i.u-tokyo.ac.jp

Moray McLaren Hewlett-Packard Labs, Palo Alto, Bristol, UK, moray.mclaren@hp.com

Matteo Monchiero Hewlett-Packard Labs, Palo Alto, CA, USA, matteo.monchiero@hp.com

Srinivasan Murali LSI, EPFL and iNoCs, Lausanne, Switzerland, murali@inocs.com

Naveen Muralimanohar Hewlett-Packard Labs, Palo Alto, CA, USA, naveen.muralimanohar@hp.com

Hiroshi Nakamura The University of Tokyo, 7-3-1 Hongo, Bunkyo-ku, Tokyo, Japan 113-8656, nakamura@hal.ipc.i.u-tokyo.ac.jp

Maurizio Palesi Dipartimento di Ingegneria Informatica e delle Telecomunicazioni, University of Catania, Italy, mpalesi@diit.unict.it

Glenn D Reinman Computer Science Department, University of California, Los Angeles, 4731-D Boelter Hall, Los Angeles, CA 90095, reinman@cs.ucla.edu

Robert Schreiber Hewlett-Packard Labs, Palo Alto, CA, USA, rob.schreiber@hp.com

Ciprian Seiculescu LSI, EPFL, Lausanne, Switzerland, ciprian.seiculescu@epfl.ch

Eran Socher School of Electrical Engineering - Physical Electronics, Tel Aviv University, 234 Wolfson EE Lab Bldg, Tel Aviv University, Ramat Aviv, Tel Aviv 69978, Israel, socher@eng.tau.ac.il

Sai-Wang Tam Electrical Engineering Department, University of California, Los Angeles, Engineering IV Building, Los Angeles, CA 90095, USA, roccotam@ee.ucla.edu

Dana Vantrease Hewlett-Packard Labs, Palo Alto, CA, USA, vantrease@hp.com

Isask'har Walter Electrical Engineering Department, Technion - Israel Institute of Technology, Haifa 32000, Israel, zigi@tx.technion.ac.il

David Wolpert University of Rochester, Rochester, NY 14627, USA, ampadu@ece.rochester.edu

Alex Yakovlev Asynchronous Systems Laboratory, School of EECE, Newcastle University, Newcastle upon Tyne, United Kingdom, Alex.Yakovlev@ncl.ac.uk

Qiaoyan Yu University of Rochester, Rochester, NY 14627, USA, ampadu@ece.rochester.edu

Ki Hwan Yum Department of Computer Science and Engineering, Texas A&M University, College Station, TX 77843-3112, USA, yum@cse.tamu.edu

Part I
Low-Level Design Techniques

Part I

Low-Level Design Technique

Chapter 1
Hybrid Circuit/Packet Switched Network for Energy Efficient on-Chip Interconnections

Mark A. Anders, Himanshu Kaul, Ram K. Krishnamurthy, and Shekhar Y. Borkar

Abstract Network on-Chip (NoC) is an interconnect fabric to connect sub-system blocks on a chip. The NoC should provide high bandwidth and low latency, should consume low energy, and should be compact. However, all these requirements are at odds and require tradeoffs at all levels. In this chapter, we discuss issues and challenges for future NoCs with demands for high bandwidth and low energy. Next, we present details of how coupling packet-switched arbitration with circuit-switched data transfer can achieve these goals. In this hybrid network, packet-switched arbitration is used to reserve future circuit-switched channels for the data transfer, eliminating the performance bottlenecks associated with pure circuit-switched networks while maintaining their power advantage. Furthermore, proximity-based data streaming increases network throughput and improves energy efficiency. Measurements of this NoC in 45 nm CMOS are described to analyze design tradeoffs.

1.1 Network on-Chip: Past, Present, and the Future

Network on-Chip (NoC) has evolved from the good old supercomputer days where computers in a cabinet, as well as multiple cabinets, were connected together to form a complete parallel computer system. These networks were primitive indeed, such as simple Ethernet at times, nevertheless sufficient to provide the necessary bandwidth with acceptable latencies. That was then, and now, with technology scaling over several generations, you can afford to have several computers themselves on a single die, connected together by a network forming a homogeneous many-core parallel computer system. To take it even further, the integration capacity is now so vast that it is possible to integrate diverse functional blocks on a chip, to be connected by a communication network, to form a heterogeneous system, what we call a system on-chip or SoC. And the network that connects these functional blocks together is the backbone of such a system. In this chapter, we discuss the state of the art in this

M.A. Anders (✉)
Intel Corporation, Hillsboro, OR, USA
e-mail: mark.a.anders@intel.com

C. Silvano et al. (eds.), *Low Power Networks-on-Chip*,
DOI 10.1007/978-1-4419-6911-8_1, © Springer Science+Business Media, LLC 2011

field, the issues and challenges that we will face in the future, and some of the prominent work addressing these issues. We also propose a hybrid packet/circuit-switched network that combines network advantages, higher resource utilization of packet-switched networks, and low power consumption of circuit-switched networks, to improve the energy-efficiency. The energy-efficiency advantage and design trade-offs will be quantified with silicon measurements of an 8 × 8 mesh NoC in 45 nm CMOS.

1.1.1 State of the Art in NoCs

Evolution of the NoC occurred over the last 3 decades, from early days of single chip microcontrollers incorporating several simple functional blocks, to today's sophisticated SoCs integrating diverse functional blocks on a chip. The early NoCs were good enough for the purpose, and as the bandwidth demand increased, they morphed into even sophisticated networks, with higher order topologies implemented with complex switches. Let us examine the evolution, comparing and contrasting their benefits as they evolved.

1.1.1.1 Buses

A bus is the simplest NoC used in the early days of microcontrollers, to connect a tiny processor core to other peripherals, such as memory, timers, counters, and serial controllers. The bus was typically narrow, of the order of 8 to 16 bits wide, spanning the entire chip, connecting almost all the agents together. Such a long bus seems very slow due to large RC delays associated with a long bus, but the chip frequency was limited by the transistor performance, not the bus. The most prominent feature of the bus is its simplicity, needing a small transistor budget. On one hand, bus utilization is limited because it is shared, arbitrated by all the agents to transfer data. On the other hand, such a shared bus also provides the benefit of broadcast and multicast.

1.1.1.2 Rings

When transistors became faster, and the bus RC delay started to dominate the operating frequency, the obvious solution was to use repeaters in the bus to improve the delay, ultimately emerging into pipelined buses, where every repeater stage of the bus is clocked. The result is a pipelined bus, with repeated bus segments, with the clocked repeater stage at the agent itself. The result is a ring, if the two far ends are connected together [1]. The advantage of a ring is that it offers higher frequency of operation, but with potentially increased latency of a number of clocks in each hop, with average node to node latency of half the number of hops. A ring is good enough for a small number of agents; however, as the number of agents grows, the latency increases linearly.

1.1.1.3   Meshes

The latency limitation of rings resulted in a higher dimensional network such as a mesh or a torus [2]. A mesh too is a segmented bus in two dimensions, with switch at each agent, but with added complexity to route data across dimensions such as from X to Y. The advantage of a mesh network is that the average latency grows slowly (square root) with the number of agents, but adds more complexity into network protocols and implementation logic, and if not careful then could create hazardous conditions such as dead-locks. Such a network can be virtualized too, with virtual channels over physical links to further improve utilization [3].

1.1.2   Issues and Challenges for the Future

As the technology continues to scale providing abundance of transistors for integration of diverse functional blocks, how will the NoCs keep up? We will now look into the challenges of NoCs in the future.

1.1.2.1   Power and Energy

Consider an SoC on 45 nm technology, with eight agents connected on the die as shown in Fig. 1.1. Successive technology generation will double the integration capacity following Moore's Law, and expect billions of transistors, with almost 64 agents by 15 nm technology, providing terascale (Tera-ops) level performance. If the agents are connected by an 8 × 8 mesh network, then the wire segments in the mesh will be of the order of 1 mm in size.

Note that as technology scales, the number of agents on the die double, and if each agent carries a switch for a mesh network, then the energy dissipated in the switches increases proportionally. The number of wires doubles too, but the length of the wire reduces. Figure 1.2a shows estimated delay and energy of a bus, and

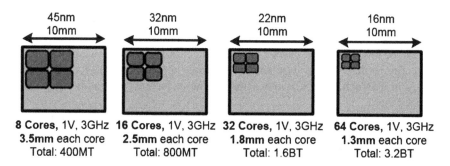

Fig. 1.1 Future integration capacity for SoCs

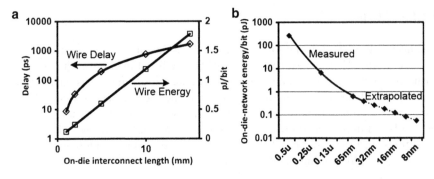

Fig. 1.2 On-die interconnect delay and energy with respect to (**a**) length and (**b**) technology

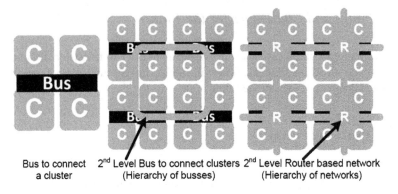

Fig. 1.3 Hierarchical, heterogeneous NoC

Fig. 1.2b shows energy expended in the switch. Using these estimates, and assuming that the Terascale SoC accesses one Tera-operand (32 bit), traversing 10 hops on an average, then the power consumption of the network alone would be too high.

1.1.2.2 Heterogeneity

Clearly, a mesh network as a homogenous NoC is not optimum. For short distances, such as adjacent agents, a bus is a much better solution because energy and delay both can be low. Moreover, buses can be designed for low voltage swings to reduce the energy further. As the wire length increases, approaching delay close to the latency in a switch, then it is more appropriate to incorporate traditional packet switched mesh.

New approaches for a NoC are needed, as shown in Fig. 1.3. Agents in close proximity could be interconnected into clusters with traditional buses which are energy efficient for data movement over short distances. The clusters could be connected together with wide (high bandwidth) low-swing (low energy) buses, or they could be connected with packet or circuit switched networks depending on the distance. Hence, the NoC could be hierarchical and heterogeneous, a radical departure from the traditional approach for NoC [4].

1.2 Proposed Hybrid Packet/Circuit Switched NoC

As integration densities continue to increase in a power-limited environment, multi-core processors provide increased performance vs. power efficiency through parallel processing at reduced voltages and frequencies. Innovations in interconnect networks for on-die communication between cores are key to enabling scalable performance as the number of cores increases [5–8]. By combining network topology and architecture advantages with efficient circuit implementations, more efficient communication can be achieved. For multi-core networks, packet-switched 2D meshes provide efficient interconnect utilization, low latencies, and high throughputs, but suffer from low energy efficiency due to data storage during routing [9,10]. Circuit-switched data transfer achieves both high bandwidth and energy efficiency by eliminating intra-route data storage [11–13]. It offers a dedicated channel during data transmission without the need for intermediate buffering or arbitration. However, by avoiding buffering and arbitration, the dedicated channel resources must be reserved prior to data transmission, possibly preventing other more optimal data transmissions from occurring. Unlike prescheduled source-directed routing schemes [2, 14], distributed routing schemes are not limited to predefined traffic patterns or applications, but determine packet routes and priorities for the reservation of resources based on incomplete real-time information. Therefore, in order to overcome challenges of resource allocation and distributed control, efficient circuits are needed that can approach throughputs of packet-switched networks while maintaining energy savings of a circuit-switched network.

1.2.1 Circuit-Switched Data with Packet-Switched Arbitration NoC

A circuit-switched 2D mesh network with packet-switched arbitration is composed of a packet-switched request address network alongside a circuit-switched acknowledge and data network (Fig. 1.4). This heterogeneous network allows delaying channel allocation for improved resource utilization, since small packets reserve channels before data transfer. However, since the data transfer uses circuit-switched paths without intermediate data storage, energy savings are also maintained. Furthermore, efficient circuits improve the overall network efficiency by reducing arbitration overhead and increasing overall utilization of network resources.

A data transmission using this hybrid network is composed of three separate phases (Fig. 1.5). During the setup phase for a circuit-switched data transmission, request packets containing the destination address are routed using the packet-

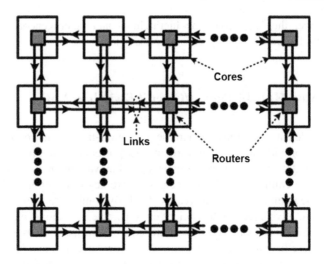

Fig. 1.4 Circuit-switched 2D mesh organization

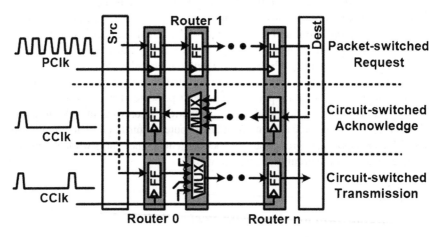

Fig. 1.5 Circuit-switched pipeline and clocking

switched network. As the request packet passes each router and interconnect segment, the corresponding circuit-switched data channel for that segment is allocated for the future circuit-switched data transmission. When the request packet reaches its destination, a complete channel or circuit has been allocated. This channel has a latching or storage element only at the destination, with only multiplexers and repeaters along the way. Acknowledge signals indicate that the channel is ready for the data transfer, thus completing the setup phase. When the channel is ready, the source router drives the data onto its output, where it propagates to the destination without interruption by state elements. Following the reception of data at the destination, the channel is deallocated at the end of the cycle.

Compared to purely packet-based networks, energy is reduced by not storing data between the source and destination. Also, since only a single header packet is transmitted to allocate each channel without multiple subsequent data flits, the traffic on the packet-switched network is also reduced. During this allocation phase, packets only hold resources at their current router while storing their routing direction for future data transfer. In contrast, a purely circuit-switched network would hold resources between the source router and its current arbitration location even when blocked by other traffic. Because circuit-switched resource allocation is a distributed optimization without global control, efficient circuits such as pipelining and queue slots can further improve overall energy efficiency by increasing utilization of the wide circuit-switched data buses. Pipelining of the three routing phases improves the data throughput. Different clocks are used to synchronize the request packet-switched and data circuit-switched portions of the network (Fig. 1.5). Since each request packet travels only between neighboring cores of each cycle, it can operate with a higher frequency clock (*PClk*) than the circuit-switched portion (*CClk*), where data may travel across the whole network of each cycle. During circuit-switched data transmissions, acknowledges for future transmissions are sent (Fig. 1.6). Also, request packets are simultaneously creating new channels by storing the routing direction for future data transmissions. This pipelining removes the request and acknowledge phases from the critical path, improving circuit-switched throughput by 3×.

In order to further improve resource utilization with distributed control, queue slots added to each router port store multiple request paths. This provides several potential paths for the circuit-switched network to choose from during the acknowledge phase. With this increase in available data transfer paths, more optimal non-interfering simultaneous data transfers occur, improving total throughput and resource utilization.

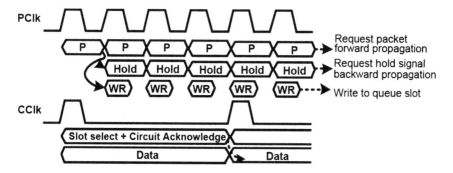

Fig. 1.6 Network timing diagram

1.2.2 Circuit Innovations for Circuit/Packet Switched Network Arbitration

Each router within the network is divided into five separate ports: north, south, east, west, and core (Fig. 1.7). Each of these ports is further divided into IN and OUT ports, for receiving and transmitting data, respectively. All ports within a router are fully connected as a crossbar. In order to avoid deadlocks, the 2D mesh uses x-first, y-second routing, and unused paths are removed from within the router. Request packets for initial arbitration are sent between neighboring routers with packet hold signals providing flow control. Bidirectional acknowledge signals, from source (*SrcAck*) and destination (*DestAck*), indicate that a circuit-switched path is ready for data transfer during the next *CClk* cycle, completing arbitration. Finally circuit-switched data is routed from source to destination.

As each request packet propagates from one router to the next, its routing direction is stored in a queue slot. During a *CClk* cycle, each router port independently selects one of its queue slots, based on a rotating priority (Fig. 1.8a). The direction previously stored in that queue slot is used to route source and destination acknowledge signals. Arrival of both acknowledges at any router along the path indicates that the complete path is ready for data transmission in the next cycle (Fig. 1.8b). Paths that are not ready must wait for a future *CClk* cycle, while ready paths free their resources following data transmission. The request packet circuits route packets containing destination address and queue slot (Fig. 1.9). The IN port compares the router and destination addresses to determine the routing direction and corresponding OUT port. Round-robin priority circuits select one of the valid packets at each OUT port and send *Hold* signals to the unselected IN ports. As each request

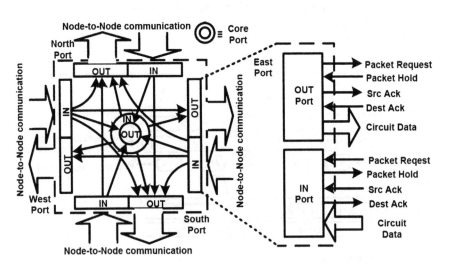

Fig. 1.7 Circuit-switched network router organization

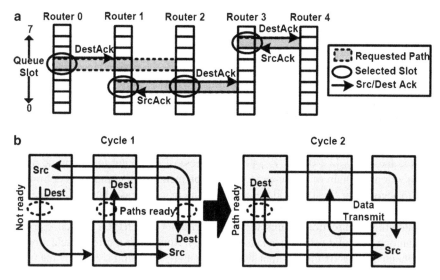

Fig. 1.8 (**a**) Slot selection and (**b**) path selection and data transmission

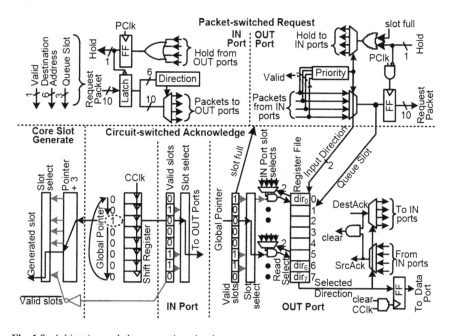

Fig. 1.9 Arbitration and slot generation circuits

packet is transmitted, the routing direction is written to the queue slot entry within a 2b register file, creating request paths from router to router. The *Hold* signal is also asserted when the requested queue slot is full or a *Hold* signal arrives from the following router, preventing that request packet's IN port from continuing.

The *Hold* signal to the previous router is delayed by one cycle using a flip-flop, reducing the *PClk* cycle time by one interconnect traversal. To accommodate the delayed *Hold* signal, an additional latch at the request packet IN port is closed whenever a *Hold* is asserted, while the previous router sends the next request packet. This results in the current packet remaining within the router, while the next packet occupies the interconnect. After the one cycle delay, the previous router will also stop transmitting. Pipelining of the flow control results in 30% *PClk* cycle time reduction. Direction circuits for the IN port coming from a core are shown in Fig. 1.10a. Each IN port compares the $X_{dest}[2:0]$ and $Y_{dest}[2:0]$ fields of a request packet with the fixed $X_{core}[2:0]$ and $Y_{core}[2:0]$ address of the router to determine routing direction. A chain of ripple-carry gates implementing 3b compare is optimally sized to account for fixed core addresses, resulting in worst-case delay of an inverter followed by two gates (Fig. 1.10b).

Following the direction circuits, the priority circuits in each OUT port choose one from among the valid request packets (Fig. 1.11). Round-robin priority selection is implemented using six circuits that select earliest arriving valid signals by comparing all pairs of valid signals in parallel, yielding 50% delay reduction compared to a tree implementation [15]. Since the circuits hold state when multiple *valid* signals arrive, deasserting the selected *valid* following transmit allows the next request packet to proceed.

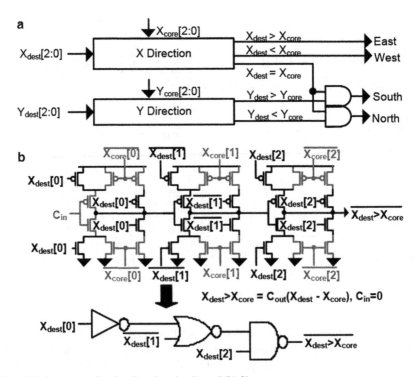

Fig. 1.10 Request packet (**a**) direction circuits and (**b**) 3b comparator

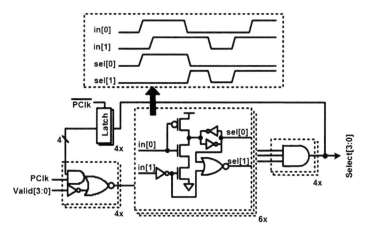

Fig. 1.11 Request packet round-robin priority circuits

Each *CClk* cycle, bidirectional source, and destination acknowledges are sent across the network. A global pointer provides the starting point for a valid queue slot selection. Rotating this pointer in each cycle using a shift register improves fairness and prevents starvation. Use of a global pointer across all routers ensures that common paths are selected. The direction information stored in the selected queue slot sets the routing direction for the two acknowledges. The interface between the request packet circuits and circuit-switched routing is shown in Fig. 1.9. All register file storage is static latch-based with interrupted feedback for robust operation. At each *PClk*, the 2b direction is written to latches addressed by the 3b queue slot. At the same time, a separate latch is set to indicate to the request packet network that the slot is now full. In each *CClk* cycle, the *full* bits are transferred to a second latch indicating *valid* queue slots. The selected queue slot is cleared if both source and destination acknowledges are asserted, indicating that data will be transferred in the following cycle. The slot select circuit searches for the next valid queue slot starting at the pointer position (Fig. 1.12). This operation is similar to an adder carry chain in which the pointer generates a one that propagates through the chain as long as the queue slots are empty. Using a logarithmic carry tree with intermediate carries that wrap around from MSB to LSB provides this functionality after three logic gates, a 63% delay reduction. A leading one from the carry merge tree indicates the position of the one-hot slot select.

1.2.3   Data Transmission Circuit Innovations

During circuit-switched data transfer, data bits are routed through multiplexers and across interconnect and repeaters from the source to the destination. Selecting new paths in each *CClk* cycle may change the routing direction at each router, causing

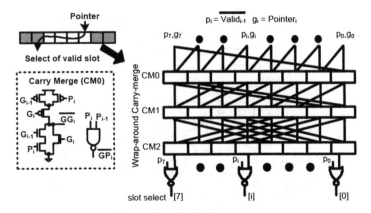

Fig. 1.12 Slot select circuit

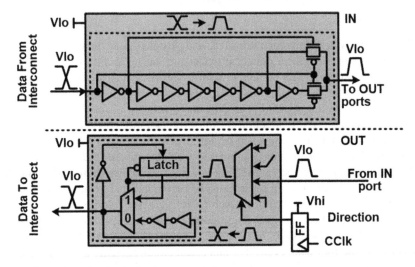

Fig. 1.13 Dual-supply circuits for data transmission

unnecessary interconnect transitions. Propagation of multiple transitions along a path significantly increases power consumption. Converting transitions to pulses through the multiplexer avoids these glitches by ensuring a common low value at all inputs when selecting new paths (Fig. 1.13). This pulse is then converted back to a transition before driving the next interconnect segment to reduce switching power on global interconnects. This reduces data switching power by up to 30%. In addition to reducing power by eliminating extra transitions, dual-supply operation (using Vlo and Vhi voltage levels), with only data transmission circuits at the lower supply, reduces power by 28% at iso-throughput while avoiding Vlo-to-Vhi level-conversion boundaries within control circuits.

Fig. 1.14 Proximity-based maximum streaming transfers per cycle

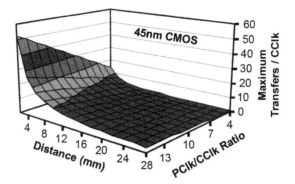

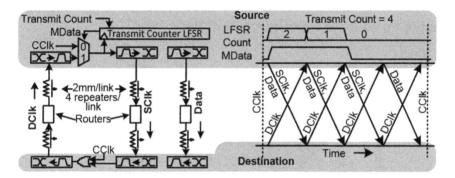

Fig. 1.15 Circuits for proximity-based streaming

Since the minimum *CClk* period is determined by opposite corner mesh traversal, data sent over shorter distances arrive at destinations early. Streaming circuits exploit this slack, allowing multiple data transfers along the same channel during a *CClk* cycle (Fig. 1.14). *SClk* and *DClk* signals, clocks from both the source and destination that accompany the data, are routed in opposite directions along the path of the data; both are initially triggered from *CClk*. *SClk* arrival at the destination triggers both data capture and the next *DClk*. Arrival of *DClk* at the source triggers data transmit and the next *SClk* if more data are available. A transmit counter at the source, loaded on *CClk* and decremented every *DClk*, provides the gating *MData* signal, indicating that more data are available (Fig. 1.15). For random data transmissions, streaming increases total network throughput up to 4.5×.

1.3 NoC Measurements and Tradeoffs in 45 nm CMOS

An 8 × 8 mesh hybrid circuit-switched network-on-chip, consisting of packet-switched arbitration logic for 512b data width with 1b circuit-switched data interconnect, is fabricated in 45 nm High-K/Metal-gate CMOS (Fig. 1.16) [16]. Each

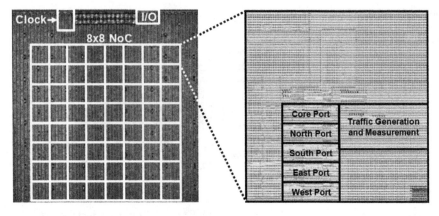

Fig. 1.16 Network-on-chip die micrograph in 45 nm CMOS

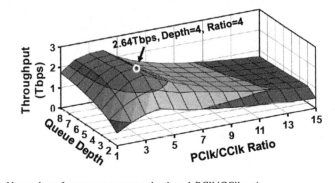

Fig. 1.17 Network performance vs. queue depth and *PClk/CClk* ratio

2 mm interconnect is folded 8× to reduce area of the prototype. On-die integrated traffic generation and measurement circuits enable static or random destination addresses with per-router programmable data rates. Using a separate supply, measured 1b data interconnect power and throughput are scaled to 512b width with 50% switching activity resulting in 560 Tbps/W energy efficiency, 4.1 Tbps bisection bandwidth, and 11 ns diagonal corner-to-corner fall-through latency. The packet-switched requests propagate at a maximum measured frequency (*PClk*) of 2 GHz, while the maximum 512 MHz *CClk* allows single-cycle opposite corner circuit-switched communication. The network achieves a maximum 2.64 Tbps throughput for random data transmissions at the optimal *PClk* to *CClk* ratio of 4 at 1.1 V, 50°C (Fig. 1.17). Lowering this ratio by reducing the *PClk* frequency reduces available paths during the acknowledge phase, while increasing the clock ratio by decreasing *CClk* frequency slows the circuit-switched transmit rate. For a 64 core network, four queue slots provide 87% throughput increase with diminishing returns for additional slots. Link utilization measures average circuit-switched interconnect use to indicate

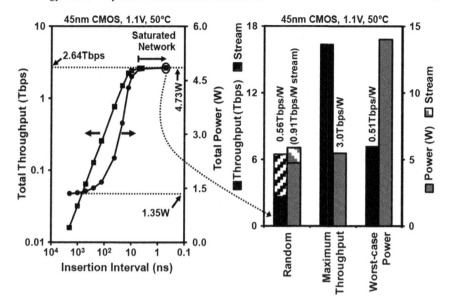

Fig. 1.18   Network throughput and power vs. data insertion rate per core and traffic

routing efficiency. Utilization of total interconnects averages 20%, while utilization of interconnects with available paths averages 50%.

Network traffic patterns vary according to destinations, whether random or static, and data insertion interval or transmit rate. Saturation of the network is achieved by increasing the rate at which data transmitted to random destinations by each of the 64 cores. The maximum throughput reaches 2.64 Tbps while average latencies increase from 8 ns to 30 ns (Fig. 1.18). Total network power peaks at 4.73 W with 0.97 W active leakage, or 74 mW/core. This power scales as network activity decreases, reaching 1.35 W or 21 mW/core with longer insertion intervals (1.1 V, 50°C). While transmissions to random destinations indicate average network performance, static traffic patterns indicate saturated network efficiency limits at maximum throughput or worst-case power (Fig. 1.19). A static traffic pattern with transmissions between neighboring Cores results in maximum throughput, increasing energy efficiency to 3.0 Tbps/W. A static pattern for worst-case power, using 88% of available circuit-switched interconnect, reduces energy efficiency to 0.51 Tbps/W.

Lowering the supply voltage decreases power consumption faster than performance, improving energy efficiency. Figure 1.20 shows the impact of supply voltage scaling on the throughput, power, and energy efficiency of a saturated network with random traffic. The energy efficiency increases from the nominal 0.56 to 1.51 Tbps/W, while operating at 634 Gbps, 420 mW measured at 550 mV, 50°C. During nominal 1.1 V operation, 83% of the total power is consumed in actual data transmission with the arbitration overhead falling to less than 10% at 550 mV.

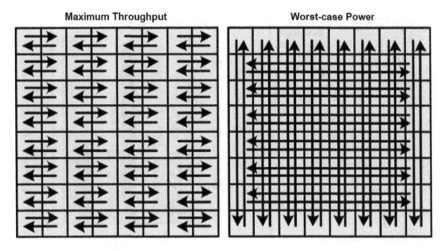

Fig. 1.19 Static traffic patterns

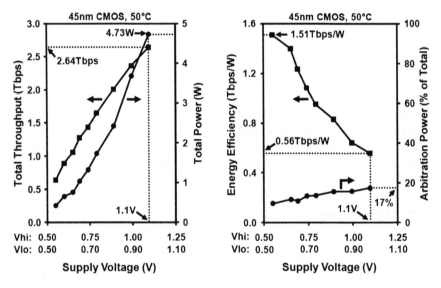

Fig. 1.20 Network throughput, power, energy efficiency, and arbitration overhead vs. supply voltage

Additionally, energy efficiency of the saturated network with random destinations can also be increased by enabling streaming transfers. Streaming circuits allow multiple data transfers along a channel once the arbitration phases are complete. As this limit on maximum transfers during a *CClk* cycle is increased, the total network throughput increases by 4.5 times to 11.8 Tbps (Fig. 1.21). Since the arbitration power is amortized across larger data transfers, energy efficiency also improves from 0.56 to 1.03 Tb/s/W.

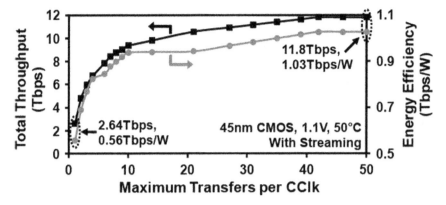

Fig. 1.21 Network performance and energy efficiency vs. streaming data transfers

1.4 Conclusion

NoC will be the backbone of the future SoC designs. These networks have evolved from simple buses to complex high dimensional networks of today, but will be limited by power and energy. We must consider hierarchical, heterogeneous networks, including packet switched and circuit switched networks, to fulfill the requirements of future SoCs. By leveraging the advantages of each network topology and switching style, hybrid networks can achieve higher energy efficiencies. The proposed hybrid circuit-switched network uses a packet network to reserve channels for future data transfer. This enables improved utilization of the circuit-switched resources, while eliminating intra-route data storage reduces overall power. Future SoCs will require multiple networks for different bandwidth, latency, energy efficiency, and area constraints that can benefit from combining the inherent advantages of different types of networks.

References

1. Pham D et al (2005) The design and implementation of a first-generation CELL processor. International Solid State Circuits Conference 49–52
2. Vangal S et al (2008) An 80-Tile Sub-100-W TeraFLOPS processor in 65-nm CMOS. IEEE Journal of Solid-State Circuits 43(1): 29–41
3. Borkar S et al (1988) iWarp: An integrated solution to high-speed parallel computing. Proceedings of Supercomputing '88 330–339
4. Borkar S (2006) Networks for multi-core Chip—A controversial view. Workshop on On- and Off-Chip Interconnection Networks for Multicore Systems (OCIN06)
5. Benini L, Micheli G (2002) Networks on chips: A new SoC paradigm. Computer Magazine 35(1): 70–78
6. Dally WJ, Towles B (2001) Route packets, not wires: On-chip interconnection networks. Design Automation Conference 684–689

7. Yu Z et al (2006) An asynchronous array of simple processors for DSP applications. International Solid State Circuits Conference 428–429

8. Keckler S et al (2003) A wire-delay scalable microprocessor architecture for high performance systems. International Solid State Circuits Conference 168–169

9. Bell S et al (2008) TILE64 – processor: A 64-Core SoC with mesh interconnect. International Solid State Circuits Conference 88–89

10. Taylor M et al (2003) A 16-issue multiple-program-counter microprocessor with point-to-point scalar operand network. International Solid State Circuits Conference 170–171

11. Wolkotte P et al (2005) An energy-efficient reconfigurable circuit-switched network-on-chip. International Parallel and Distibuted Processing Symposium 155a

12. Anders M et al (2008) A 2.9Tb/s 8W 64-Core circuit-switched network-on-chip in 45nm CMOS. European Solid State Circuits Conference 182–185

13. Anders M et al (2010) A 4.1Tb/s bisection bandwidth 560Gb/s/W streaming circuit-switched 8x8 mesh network-on-chip in 45nm CMOS. International Solid State Circuits Conference 110–111

14. Wu CM, Chi HC (2005) Design of a high-performance switch for circuit-switched on-chip networks. Asian Solid States Circuits Conference 481–484

15. Lee K. et al (2003) A high-speed and lightweight on-chip crossbar switch scheduler for on-chip interconnection networks. European Solid State Circuits Conference 453–456

16. Mistry K et al (2007) A 45nm logic technology with high-k+Metal gate transistors, strained silicon, 9 Cu interconnect layers, 193nm dry patterning, and 100% Pb-free packaging. International Electron Devices Meeting 247–250

Chapter 2
Run-Time Power-Gating Techniques for Low-Power On-Chip Networks

Hiroki Matsutani, Michihiro Koibuchi, Hiroshi Nakamura, and Hideharu Amano

Abstract Leakage power has already been consuming a considerable portion of the active power in recent process technologies. In this chapter, we survey various power gating techniques to reduce the leakage power of on-chip routers. Then we introduce a run-time fine-grained power-gating router, in which power supply to each router component (e.g., virtual-channel buffer, crossbar's multiplexer, and output latch) can be individually controlled in response to the applied workload. The fine-grained power gating router with 35 micro-power domains is designed using a commercial 65 nm process and evaluated in terms of the area overhead, application performance, and leakage power reduction.

2.1 Introduction

Network-on-Chips (NoCs) have been used not only in high-performance microarchitectures, but also in cost-effective embedded devices mostly used in consumer equipments, such as set-top boxes or mobile wireless devices. These embedded applications usually require low power, since power consumption is the dominant factor on their battery life, heat dissipation, and packaging cost.

The overall power consumption consists of dynamic switching power and static leakage power. Switching power is still the major component of the overall power consumption during active operations. In addition, we need to take care of the leakage power, since it has already been consuming a considerable portion of the active power in recent process technologies, and it will further increase while switching power becomes smaller when the technology is scaled down. Different saving techniques have been used for reducing each type of power. For example, clock gating, operand isolation, and dynamic voltage and frequency scaling (DVFS) target on switching power reduction. The design using transistors with multi-threshold voltages including power gating has been used for leakage power reduction.

H. Matsutani (✉)
The University of Tokyo, 7-3-1 Hongo, Bunkyo-ku, Tokyo, Japan 113-8656
e-mail: matsutani@hal.ipc.i.u-tokyo.ac.jp

C. Silvano et al. (eds.), *Low Power Networks-on-Chip*,
DOI 10.1007/978-1-4419-6911-8_2, © Springer Science+Business Media, LLC 2011

We mainly focus on the power gating to reduce the leakage power of NoC, since it is consumed without any packet transfers as long as the NoC is powered on. As the NoC is the communication infrastructure of various SoCs, it must be always ready for the packet transfers at any workload so as not to increase the communication latency; thus, we have studied some run-time power management techniques that dynamically stop the leakage current whenever possible [16, 17].

In this chapter, we introduce a run-time fine-grained power-gating router, in which power supply to each router component (e.g., virtual-channel buffer, crossbar's multiplexer, and output latch) can be individually controlled in response to the applied workload. As only the router components along an active datapath which is just transferring a packet are activated, the leakage power of the on-chip network can be reduced to the near-optimal level. However, such run-time power gating inherently increases the communication latency and degrades the application performance, since a certain amount of wakeup latency is required to activate the sleeping components. To mitigate the wakeup latency, we introduce three early wakeup methods that can detect the next packet arrival and activate the corresponding components in advance. The fine-grained power-gating router with the early wakeup methods is evaluated in terms of the application performance, area overhead, and leakage power.

The rest of this chapter is organized as follows. Section 2.2 shows architecture of a typical on-chip router and analyzes its power consumption. Section 2.3 surveys low-power techniques especially power gating for NoCs. Section 2.4 designs the fine-grained power-gating router and Sect. 2.5 proposes three early wakeup methods. Section 2.6 evaluates the power gating router with the early wakeup methods. Finally, Sect. 2.7 summarizes this chapter.

2.2 On-Chip Virtual-Channel Router

Before discussing low-power techniques for on-chip routers, an architecture of a simple on-chip virtual-channel (VC) router is presented, and then its dynamic and static power consumption is analyzed.

2.2.1 Target Router Architecture

For investigation on NoC architectures, we have implemented a wormhole router with virtual channels. We also developed an NoC generator that automatically connects the routers in arbitrary network topologies. The generated NoC is synthesized, placed, and routed with a 65 nm standard cell library.

Figure 2.1 illustrates the virtual-channel router. The router has p input and output physical channels, a $p \times p$ crossbar switch, and a round-robin arbiter that allocates a pair of output virtual and physical channels for each incoming packet. Each input

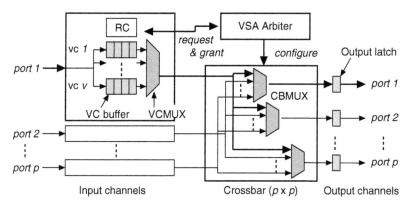

Fig. 2.1 Virtual-channel router architecture used

physical channel has a separated buffer queue for each virtual channel, while each output physical channel has a single one-flit buffer (**Output latch**) to decouple the switch and link delays.

To overcome the head-of-line blocking, a virtual-channel design sometimes uses a $pv \times pv$ full crossbar, where p is the number of physical channels and v is the number of virtual channels. However, the crossbar complexity is significantly increased with the $pv \times pv$ crossbar, and its performance improvement will be limited because the data rate out of each input port is limited by its bandwidth [5]. Therefore, we used a small $p \times p$ crossbar by just duplicating the buffers. It is composed of p p-to-1 multiplexers (**CBMUXes**), each of which is controlled by a select signal from the arbiter.

Each input physical channel has v virtual channels. It has a routing computation (RC) unit and a v-to-1 multiplexer (**VCMUX**) that selects only a single output from v virtual channels. Each virtual channel has a control logic, status registers, and an n-flit FIFO buffer (**VC buffer**). The RC unit in this design is very simple, because routing decisions are stored in the header flit prior to packet injection (i.e., source routing); thus, routing tables that require register files for storing routing paths are not needed.

The router architecture is fully pipelined. Although some one- or two-cycle routers have been developed by using some aggressive techniques [18, 19], we selected a simple three-cycle router architecture illustrated in Fig. 2.2. In this figure, a packet that consists of a single header flit and three body flits is transferred from Router A to Router C. Each router transfers a header flit through three pipeline stages that consist of routing computation (RC) stage, virtual channel and switch allocation (VSA) stage, and switch traversal (ST) stage.

Finally, we designed an RTL model of the router. Its parameters p, v, and n are set to 5, 4, and 4, respectively. The flit width w is set to 128-bit. These FIFO buffers can be implemented with either SRAMs or flip-flops (FFs), depending on the depth of the buffers, not the width. We assume that buffers should be implemented with

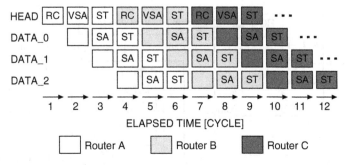

Fig. 2.2 Three-stage router pipeline. A packet is transferred from router A to C

macro cells if their depths are more than 16. Otherwise buffers should simply be implemented with FFs. Since the depth of the FIFO buffers in this design is only four, the input buffers are implemented with small FFs.

2.2.2 Power Analysis of the Target Router Architecture

To estimate the power consumption of the router mentioned previously, the following steps are performed:

1. The RTL model of the router is synthesized by the Synopsys Design Compiler.
2. The netlist is placed and routed (including a clock tree synthesis and buffer insertion) by the Synopsys Astro.
3. The placed-and-routed design is simulated by the Cadence NC-Verilog, to obtain the switching activity information of the router.
4. The power consumption is estimated from the switching activity by the Synopsys Power Compiler.

A 65 nm CMOS process with a core voltage of 1.20 V is selected for this analysis. Clock gating and operand isolation are fully applied to the router to minimize its switching activity and dynamic power.

In step 3, the router is simulated at 500 MHz with various fixed workloads (i.e., throughputs), in the same manner as in [2]. A packet stream is defined as intermittent injections of packets that use approximately 30% of the maximum link bandwidth of a single router link. Each header flit contains a fixed destination address, while the data flits contain random values as a payload. The number of packet streams injected into the router is changed so as to generate various workloads. In this experiment, up to five streams are applied to the five-port router, and the power consumption at each workload level is analyzed.

Figure 2.3 shows power consumption of the router with various workloads and temperatures. The router consumes more power as it processes more packet streams, in the following way:

$$P_{\text{total}} = P_{\text{standby}} + x P_{\text{stream}}, \tag{2.1}$$

where x is the number of packet streams and P_{stream} is the dynamic power for processing a packet stream.

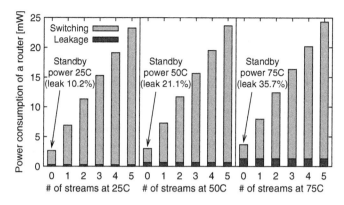

Fig. 2.3 Power consumption of router with various workloads (temperature: 25°C, 50°C, and 75°C)

Note that the router consumed a certain amount of power even with no traffic (i.e., $P_{standby}$). Leakage power consumes a substantial portion of the standby power, and it drastically increases as the temperature increases. For example, the leakage power consumes 35.7% of the standby power when the temperature is 75°C. The remaining standby power is the dynamic power consumed by the clock tree buffers and the latches inserted for the clock gating; hence, further reduction of the switching activity would be difficult.

The leakage power is consumed without any packet transfers as long as the NoC is powered on; thus it cannot be negligible. In this chapter, we introduce power-gating techniques to reduce the leakage power.

2.3   Previous Work on Low-Power Techniques

Various low-power techniques have been used for microprocessors and on-chip routers. In particular, clock gating and operand isolation are common techniques and they have already been applied to our router design. Section 2.3.1 surveys the voltage and frequency scaling techniques. Section 2.3.2 surveys the run-time power gating and network link shutdown techniques.

2.3.1   Voltage and Frequency Scaling Techniques

The dynamic voltage and frequency scaling (DVFS) is a power-saving technique that reduces the operating frequency and supply voltage according to the applied workload. The dynamic power consumption is related to the square of the supply voltage; thus, because a peak performance is not always required during the whole execution time in most cases, adjusting the frequency and supply voltage so as to at least achieve the required performance can reduce the dynamic power.

DVFS has been applied to various circuits, such as microprocessors, accelerators, and network links. In [21], the frequency and the voltage of network links are dynamically adjusted based on the past utilization. In [23], the network link voltage is scaled down by distributing the traffic load using an adaptive routing. The frequency is typically controlled by a PLL frequency divider, and the supply voltage can be adjusted by controlling an off-chip dc–dc converter.

2.3.2 Power Gating Techniques

Power gating is a representative leakage-power reduction technique, which shuts off the power supply of idle circuits blocks by turning off (or on) the power switches, which are inserted between the GND line and the blocks or between the VDD line and the blocks. Leakage current can be much reduced by the power switch made with high threshold low leakage transistors without reducing the speed of the target circuits block using low-threshold high-speed transistors. This concept has been applied to circuits blocks with various granularities, such as processor cores or IP modules [10, 11], execution units in a processor [7, 9, 20], and primitive gates. Depending on the granularity of target circuits blocks (i.e., power domains), the power gating is classified into coarse-grained approach and fine-grained one.

2.3.2.1 Coarse-Grained Power-Gating Techniques

In the coarse-grained approach, a target circuits block is surrounded by a power/-ground ring. Power switches are inserted between the core ring and power/ground IO cells. The power supply to the circuits block can be controlled by the power switches. Since the power supply to all cells inside the core ring is controlled at one time, this approach is well suited to the IP- or module-level power management. The coarse-grained approach has been popularly used, since its IP- or module-level power management is straightforward and easy to control.

2.3.2.2 Fine-Grained Power Gating Techniques

The fine-grained approach has received a lot of attention in recent years because of its flexibility and short wakeup latency [8, 9, 20, 24].

Although various types of fine-grained power-gating techniques have been proposed, we focus on the method proposed in [24]. In this method, customized standard cells, each of which has a virtual ground (VGND) port by expanding the original cell, are used. These standard cells that share the same active signal form

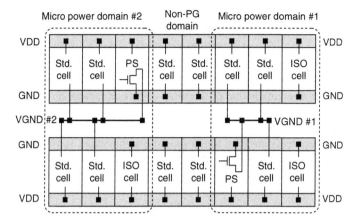

Fig. 2.4 Fine-grained power gating. Two micro power domains and a single non power-gating domain are illustrated. PS and ISO refer to a power switch and an isolation cell, respectively

a single micro-power domain, by connecting their VGND ports to a shared local VGND line, as shown in Fig. 2.4. Power switches are inserted between the VGND line and GND line for controlling the power supply to the micro-power domain. Figure 2.4 illustrates two micro-power domains and a single non-power-gating domain. Each micro-power domain has its own local VGND line and power switch.

Compared with the UPF-based methodology [8], it can control the number of power switches more flexibly by considering a given wakeup latency requirement. That is, various shapes of micro-power domains can be formed by grouping the standard cells that share the same active signal. The wakeup latency of this method is typically less than a few nanoseconds (we will confirm this in Sect. 2.4.3). A real chip implementation of a microprocessor which uses this fine-grained power gating has been reported recently [9] and its feasibility has also been confirmed.

In this chapter, we only focus on fine-grained approach because coarse-grained one is difficult to be used for on-chip routers. Each input physical channel works independently of each other unless packet contentions with the other physical channels occur. In addition, all virtual channels in the same physical channel are not always used. Actually zero or only a few virtual channels are occupied at the same time in most cases. This indicates that a finer-grained partitioning can exploit the spatial and temporal communication locality and increase the power-gating opportunities.

2.3.2.3 Power Gating for Interconnection Networks

As the standby power consumption is becoming more and more serious, various power-gating techniques have been applied to on-chip routers to reduce the standby leakage power [16, 17, 25]. In [25], each router is partitioned into ten smaller sleep regions with control of individual router ports. An input physical channel level power gating is studied in [17], while a virtual channel level power management is discussed in [16]. In [3], PMOS power switches controlled by an ultra-cut-off (UCO) technique are inserted in each NoC unit to maintain minimum leakage in standby mode.

When the power-gating techniques are applied to on-chip networks, the wakeup control of power domains is one of the most important factors on the application performance. In [22], the authors provide a thorough discussion about power-aware networks whose links can be turned on and off, in terms of connectivity, routing, wake-up and sleep decisions, and router pipeline architecture. In [4], as a leakage-power aware buffer management method, a certain portion (i.e., window size) of the buffer is made active before it is accessed, in order to remove the performance penalty. By tuning the window size, the input ports can always provide enough buffer space for the arrival of packets, and the network performance will never be affected [4].

In Sect. 2.4, we introduce the fine-grained power gating router with 35 micro-power domains. In Sect. 2.5, we introduce three early wakeup methods that can mitigate or completely remove the negative impact of the wakeup latency.

2.4 Fine-Grained Power Gating Router

Here, we first show how an on-chip router is divided into a number of micro-power domains. Then we implement these power domains using a 65 nm process and evaluate them in terms of the area overhead and wakeup latency.

2.4.1 Power Domain Partitioning

Before partitioning the on-chip router into a number of micro-power domains, we should estimate the gate count of each router component, since the leakage power is proportional to the device area. The RTL model of the router designed in Sect. 2.2.1 is used. As mentioned before, the router has five input physical channels, each of which has four virtual channels. Each virtual channel has a four-flit buffer queue. The flit width is 128-bit.

Table 2.1 shows the gate count of each router component, such as VC buffer, output latch, CBMUX, and VCMUX (Fig. 2.1). In this table, "Others" include the gate counts of routing computation units, an arbiter, VC status registers, and the other control logic, but each of these components is quite small compared to the other components. Actually, these miscellaneous logics consume only 11.9% of the router area; so they are removed from the power domain list in order to simplify the power gating router design. Consequently, the router area is divided into 35 power domains including VC buffers, Output latches, VCMUXes, and CBMUXes, which can cover 88.1% of the total router area.

Table 2.1 Total gate count of each router component (before PS insertion) [kilo gates]

Module	Count	Total gate count	Ratio
4-flit VC buffer	20	111.06	77.9%
1-flit output latch	5	5.49	3.9%
5-to-1 CBMUX	5	4.91	3.4%
4-to-1 VCMUX	5	4.21	3.0%
Others	1	16.92	11.9%
Total		142.58	100%

2.4.2 Power Domain Implementation

Here, we design all power domain types (i.e., VC buffer, Output latch, CBMUX, and VCMUX) in order to estimate their area overhead and wakeup latency.

The following design flow is used for all power domain types.

1. An RTL model of a power domain with an active signal is designed.
2. The RTL model is synthesized by the Synopsys Design Compiler.
3. Isolation cells are inserted to all output ports of the synthesized netlist in order to hold the output values of the domain when the power supply is stopped.
4. The netlist with isolation cells is placed by the Synopsys Astro.
5. The virtual ground (VGND) port of each cell is connected, and the power switches are inserted between the VGND and GND lines by the Sequence Design CoolPower.
6. The netlist with power switches is routed by the Synopsys Astro.
7. The previous two steps are performed again in order to optimize the VGND, power switch sizing, and routing.

Using this design flow, we obtained layout data (GDS files) of VC buffer, Output latch, CBMUX, and VCMUX. Note that this flow is fully automated; so an additional design complexity for the fine-grained power gating is small.

Table 2.2 shows the area overhead of the isolation cells and power switches for each router component. In this table, ISO and PS show the total gate counts of isolation cells and power switches used in the router, respectively. In the Overhead column, "ISO,PS only" shows the area overhead of the isolation cells and power switches. The total area overhead of the ISO and PS cells is only 4.3%.

In this fine-grained power gating, we used the customized standard cells each of which has a VGND port. We selected 106 cells from a commercial 65 nm standard cell library and modified them to have a VGND port by expanding their cell height. In the Overhead column of Table 2.2, "+Cell height" considers the area overhead of the customized standard cells against the original ones, in addition to ISO and PS cells. In this case, the total area overhead is increased to 15.9% but it is still reasonable, since leakage current of these cells can be cut off when they are not activated.

Table 2.2 Total gate count of each router component (after PS insertion) [kilo gates]

Module	Count	ISO	PS	Overhead ISO,PS only	+Cell height
4-flit VC buffer	20	2.07	2.25	3.9%	15.4%
1-flit output latch	5	0.51	0.16	12.2%	24.6%
5-to-1 CBMUX	5	0.52	0.02	10.9%	23.3%
4-to-1 VCMUX	5	0.54	0.02	13.3%	25.9%
Others	1	0	0	0%	11.1%
Total		3.64	2.44	4.3%	15.9%

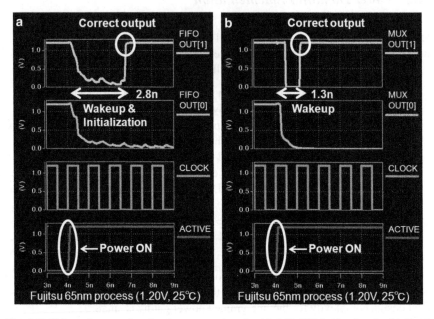

Fig. 2.5 Wakeup latency of each micro-power domain. (**a**) 4-flit VC buffer domain (**b**) 5-to-1 CBMUX domain

2.4.3 Wakeup Latency Estimation

To estimate the wakeup latency of each power domain, the following steps are performed:

1. A SPICE netlist of the target power domain is extracted from its layout data by the Cadence QRC Extraction.
2. The wakeup latency is measured based on circuit simulations of the SPICE netlist by the Synopsys HSIM.

Figure 2.5 shows the measured waveforms of the VC buffer and the CBMUX domains when their active signal is asserted. The waveforms of the Output latch and VCMUX domains are omitted, since they are quite similar to those of the VC buffer and CBMUX domains, respectively. In each figure, the first (top) and second waveforms show the lower two bits of the output (OUT[1] and OUT[0]) from the domain. The third waveform shows the 1GHz clock signal and the fourth one shows the active signal. In these simulations, the lower two bits of input (IN[1] and IN[0]) are set to 1 and 0, respectively. Then, the active signal is asserted at the second rising edge of the clock. As shown in these figures, the output values of the VC buffer reach to the expected values within 2.8 ns, while those of the CBMUX take approximately 1.3 ns.

Therefore, we assume the wakeup latency of each power domain is two, three, and four cycles when the target NoC is operated at 667 MHz, 1 GHz, and 1.33 GHz, respectively. This assumption is little bit conservative, since the actual wakeup latencies are less than 3 ns as mentioned above.

2.5   Wakeup Control Methods

In this section, we first show the negative impact of the wakeup latency on application performance. Then we introduce three wakeup control methods that mitigate the wakeup latency impact.

2.5.1   Wakeup Latency Impact

NoCs are quite latency sensitive, since their communication latency determines the application performance on many-core architectures. Thus, the wakeup latency of router components imposed by the power gating should be hidden.

To clearly show the negative impact of the wakeup latency, we preliminarily evaluate the application performance of a NoC without any early wakeup methods. As a target NoC, here we assume a NoC used in a chip multiprocessor illustrated in Fig. 2.10. Details about the chip multiprocessor will be described in Sect. 2.6.

Figure 2.6 shows the execution cycles of ten application programs selected from SPLASH-2 benchmark: (a) radix, (b) lu, (c) fft, (d) barnes, (e) ocean, (f) ray-trace, (g) volrend, (h) water-nsquared, (i) water-spatial, and (j) fmm. As wakeup latencies, two, three, and four cycles are simulated. In this graph, 1.0 indicates the original application performance without power gating (i.e., no wakeup latency). As shown in Fig. 2.6, the average execution time of these applications is increased by 23.2%, 35.3%, and 46.3% when the wakeup latency is two, three, and four cycles, respectively.

Such performance penalty is unacceptable even if the leakage power is significantly reduced, since more computation power or higher clock frequency will be required to compensate the performance penalty. To mitigate or remove the wakeup latency, an early wakeup method that can detect the next packet arrival and activate the corresponding components in advance is essential. We will introduce three early wakeup methods in the following sections.

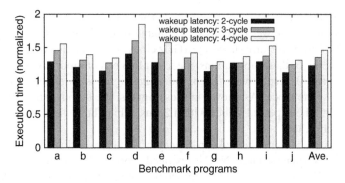

Fig. 2.6 Execution time of SPLASH-2 benchmark without early wakeup methods (1.0 indicates the execution time without power gating)

2.5.2 Look-Ahead Method

Concept. To wake up each micro-power domain as early as possible, Look-ahead method employs the look-ahead routing [6, 17] that can detect which input channel of two hops away will be used. Since an activation of a micro-power domain is triggered prior to several cycles before packets actually reach to the domain, it can mitigate or remove the negative impact of the wakeup latency.

Early detection of packet arrival. Figure 2.7 illustrates how the look-ahead routing detects which input channel of two hops away will be used. In this figure, NRC denotes the routing computation for the next hop. Assuming a packet is transferred from Router A to Router C via Router B, the NRC unit at Router A computes the output channel of the next router (i.e., Router B), instead of its own output channel. Since the output channel of Router B is directly connected to an input channel of Router C, the NRC unit at Router A can detect which output channel of Router B and which input channel of Router C will be used. As shown in Fig. 2.7, Router A can trigger the activation of Router C when it completes its NRC. There is a five cycle margin after a packet completes the NRC of Router A until the packet actually reaches to Router C. Thus, it can remove the wakeup latencies of less than five cycles.

Wake-up scheduling. Each micro-power domain (i.e., VC buffer, VCMUX, CB-MUX, and Output latch) is activated in different timing. Assuming a packet is transferred from Router A to Router C via Router B, each power domain of Router C is woken up as follows:

- **VC buffer:** Activation of an input VC buffer in Router C is triggered when a packet header in Router A completes the NRC operation.
- **VCMUX:** Activation of a VCMUX in Router C is triggered when a packet header in Router A completes the NRC operation.

Fig. 2.7 Router pipeline using Look-ahead wakeup control method

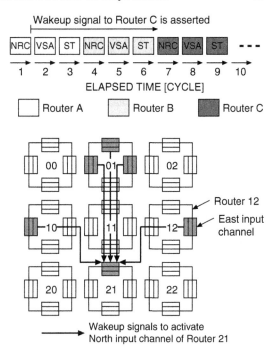

Fig. 2.8 Wakeup signals to wakeup the North input channel of router 21 in 3 × 3 mesh NoC

- **CBMUX:** Activation of a CBMUX in Router C is triggered when a packet header in Router B completes the NRC operation.
- **Output latch:** Activation of an Output latch in Router C is triggered when a packet header in Router B completes the NRC operation.

Wake-up control network. To deliver these wakeup signals, a wakeup control network is needed in the NoC. That is, an input channel (or NRC unit) of Router A has to deliver the wakeup signal to the corresponding VC buffer and VCMUX of Router C. The NRC unit of Router B has to deliver the wakeup signal to the corresponding CBMUX and Output latch of Router C. Figure 2.8 shows an example of wakeup signals in a 3 × 3 mesh network. This figure illustrates five wakeup signals to wake up the North input channel (i.e., VC buffer and VCMUX) of Router 21. In this example, North, West, and East input channels of Router 01, West input channel of Router 10, and East input channel of Router 12 have the wakeup signal to wake up the North input channel of Router 21, respectively. The target input channel is activated when one or more wakeup signals are asserted. Every input channel in the network has such wakeup signals and monitors them to be activated or deactivated.

The wakeup signal spans the twice longer distance than a wire between two neighboring routers; thus an additional cycle would be required to deliver the wakeup signal, depending on the distance between two routers.

The first hop problem. Another difficulty of Look-ahead method is the wakeup control of the first hop. We assume that the source network interface (source NI) triggers the activation of the first and second hops during the packetization. However, assuming the source NI triggers the activation of the first hop one cycle ahead, the wakeup latency of the first hop cannot be hidden. In this case, Look-ahead method compensates only one cycle of the first-hop wakeup latency but suffers the remaining wakeup latency.

Assuming the i-th hop router can mitigate T^i_{recover} cycles of the wakeup latency, T^i_{recover} is calculated as follows.

$$T^i_{\text{recover}} = \begin{cases} 2n - T_{\text{wire}} - 1 & i \geq 2 \\ 1 & i = 1 \end{cases}, \qquad (2.2)$$

where n is the router pipeline depth (e.g., three stages) and T_{wire} is the wire delay of a wakeup signal. Assuming $n = 3$ and $T_{\text{wire}} = 1$, the second or farther hop routers can mitigate up to four cycles of the wakeup latency, while the first hop mitigates only a single cycle.

2.5.3 Look-Ahead with Ever-On Method

Concept. Look-ahead with ever-on method is an extension of the original Look-ahead method to mitigate the wakeup latency of the first hop. In the ever-on method, VC buffers that are activated frequently as the first hop are selected as "ever-on" domains, and their power supplies are never stopped. No wakeup latency is required for the ever-on domains. However, the ever-on domains must be selected quite carefully, since they always consume leakage power. The other power domains are woken up in the same way as the original Look-ahead method.

Ever-on selection. To select the ever-on domains, we should analyze traffic patterns of the target NoC. In this chapter, for example, the fine-grained power gating router is applied to a chip multi-processor illustrated in Fig. 2.10. In this case, only the VC buffers that are directly connected from processor cores and used heavily should be selected as ever-on domains in order to mitigate the first-hop latency with minimum leakage power overhead. As a result, we selected VC buffers of VC0 and VC2 directly connected from the eight processor cores as ever-on domains in this study.

We will evaluate the impact of these ever-on domains in terms of the application performance and leakage power in Sect. 2.6.

2.5.4 Look-Ahead with Active Buffer Window Method

Concept. Look-ahead with active buffer window (ABW) method is another extension of the original Look-ahead method to completely remove the first-hop wakeup latency by keeping a part of each VC buffer active. That is, a part of VC buffer is

Fig. 2.9 Example of active buffer window. VC buffer is divided into flit-level domains

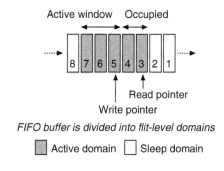

always kept active to prepare for the next flit arrivals, while such active parts always consume leakage power. This ABW method is inspired by a leakage-aware buffer management proposed in [4].

Active buffer window. In the ABW method, each VC buffer domain is divided into more finer flit-level power domains, each of which can be activated and deactivated independently. For example, Fig. 2.9 illustrates an eight-flit VC buffer, which is divided into eight flit-level power domains. To store incoming flits without any wakeup latency, a certain amount of the flit-level power domains in each VC buffer are always kept active, regardless of the workload. The activated part of the VC buffer is called "active buffer window." The flit-level domains are activated so that size of the active buffer window is constant at any time. That is, the active buffer window in a VC buffer moves whenever a part of the active buffer is consumed, in order to prepare for successive flit arrival. In Fig. 2.9, the active buffer window size is three. Two flit-level domains are already occupied by flits and the next three domains are activated to prepare for successive flits.

The ABW method is applied to only VC buffers. The other power domains are woken up in the same way as the original Look-ahead method.

2.6 Experimental Evaluations

In this section, we evaluate the fine-grained power-gating router with the early wakeup methods in terms of the application performance and leakage power.

2.6.1 Simulation Environment

An NoC used in the eight-core CMP shown in Fig. 2.10 is simulated. As shown, eight processor cores and 64 L2 cache banks are interconnected by 16 on-chip routers. The cache architecture is SNUCA [12]. Since these L2 cache banks are shared by all processors, a cache coherence protocol is running on the CMP. Table 2.3 lists the processors and memory system parameters, and Table 2.4 shows

Fig. 2.10 Target CMP
architecture in this evaluation

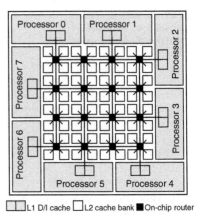

Table 2.3 Simulation
parameters (processor
and memory)

Processor	UltraSPARC-III
L1 I-cache size	16 KB (line:64B)
L1 D-cache size	16 KB (line:64B)
# of processors	8
L1 cache latency	1 cycle
L2 cache size	256 KB (assoc:4)
# of L2 cache banks	64
L2 cache latency	6 cycle
Memory size	4 GB
Memory latency	160 cycle

Table 2.4 Simulation
parameters (router
and network)

Topology	4 × 4 mesh
Routing	Dimension-order
Switching	Wormhole
# of VCs	4
Buffer size	4 flit
Router pipeline	[RC][VSA][ST]
Flit size	128 bit
Control packet	1 flit
Data packet	5 flit

the on-chip routers and network parameters. To simulate the above-mentioned CMP, we used a full-system multiprocessor simulator: GEMS [15] and Virtutech Simics [13].

Network Model: We modified a detailed network model of GEMS, called Garnet [1], in order to accurately simulate the proposed fine-grained power gating of on-chip routers and the three wakeup methods. As illustrated in Table 2.4, typical three-stage pipelined routers are used in the NoC. In the Look-ahead-based wakeup methods, the wire delay of a wakeup signal T_{wire} is set to one cycle, and the size of a VC buffer in the router is set to four flits. In the ABW method, the active buffer window size is set to two flits.

Cache Coherence Protocol: Token coherence protocol [14] is used. To avoid end-to-end protocol (i.e., request-reply) deadlocks, the on-chip network uses four virtual channels (VC0 to VC3) as follows.

- **VC0:** Request from L1 cache to L2 cache bank; request from L2 cache bank to L1 cache
- **VC1:** Request from L2 cache bank to directory controller; request from directory controller to L2 cache bank
- **VC2:** Reply from L1 cache (or directory controller) to L2 cache bank; reply from L2 cache bank to L1 cache (or directory controller)
- **VC3:** Persistent request from L1 cache

The utilization ratio of each virtual channel is not to be equal. The utilization of VC1 is low when the main memory is accessed sparsely due to frequent cache hits. VC3 is assigned to the persistent requests for avoiding the starvation. Its traffic amount is quite small since such situation is not so frequent (e.g., 0.19% of all requests [14]). The proposed fine-grained power-gating technique can exploit such imbalanced use of power domains inside a router.

Benchmark Programs: To evaluate the application performance of the proposed fine-grained power gating with different wakeup methods, we used ten parallel programs of SPLASH-2 benchmark [26]. Sun Solaris 9 operating system is working on the eight-core CMP. These benchmark programs are compiled using Sun Studio 12 and they are executed on Solaris 9. The number of threads is set to eight in each program.

2.6.2 Performance Impact

In this section, we evaluate the performance impact of the run-time power gating and its early wakeup methods.

We count the execution cycles of ten benchmark programs when the proposed fine-grained power gating technique is applied to on-chip routers in the CMP. We also compare the proposed early wakeup methods in terms of the application performance. As mentioned in Sect. 2.4.3, the wakeup latency of each power domain is less than 3 ns; thus, we assume that the wakeup latency is two, three, and four cycles in our simulations when the target NoC is operated at 667 MHz, 1 GHz, and 1.33 GHz, respectively.

Figure 2.11a shows the application performance of the original Look-ahead, Look-ahead with ever-on, and Look-ahead with ABW methods, when the wakeup latency of every domain is set to two cycles. The benchmark set includes (a) radix,

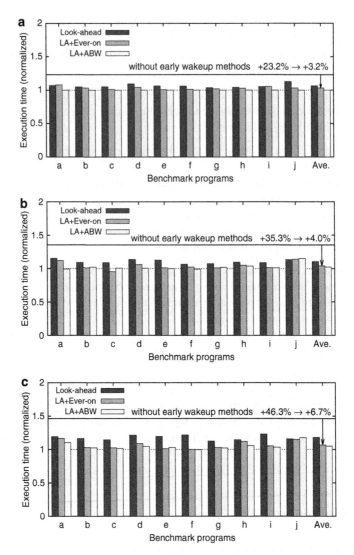

Fig. 2.11 Execution time of SPLASH-2 benchmark with different wakeup methods (1.0 indicates the execution time without power gating). (**a**) Wakeup latency: 2-cycle (**b**) Wakeup latency: 3-cycle (**c**) Wakeup latency: 4-cycle

(b) lu, (c) fft, (d) barnes, (e) ocean, (f) raytrace, (g) volrend, (h) water-nsquared, (i) water-spatial, and (j) fmm. Each application performance is normalized so that the original application performance without power gating (i.e., 0-cycle wakeup) is set to 1.0.

As shown in the graph, a similar tendency is observed in all the programs. Although the power gating with no early wakeup methods increases the execution time by 23.2% (see Sect. 2.5.1), those with the original Look-ahead, Look-ahead ever-on, and Look-ahead ABW methods increase only 6.3, 3.2, and 0.0% on average, respectively.

Figure 2.11b illustrates the application performance of the three wakeup methods when the wakeup latency is three cycles under an operating frequency of 1 GHz. In this case, the power gating with no early wakeup methods increases the execution time by 35.3%. On the other hand, those with the original Look-ahead, Look-ahead ever-on, and Look-ahead ABW methods increase 10.5, 4.0, and 2.4% on average, respectively. Thus, Look-ahead ever-on and Look-ahead ABW methods can successfully mitigate the wakeup latency when the target NoC is running at 1 GHz.

Figure 2.11c shows the application performance when the wakeup latency is four cycles under an operating frequency of 1.33 GHz. The power gating with no early wakeup methods increases the execution time by 46.3%, while those with Look-ahead ever-on and Look-ahead ABW methods increase only 6.7 and 4.9%, respectively. Thus, these early wakeup methods are still reasonable at such a high operating frequency.

2.6.3 Leakage Power Reduction

In this section, we estimate the average leakage power of the proposed power-gating router with different wakeup methods when the application workload is applied.

Table 2.5 shows leakage power of each router component, based on the post-layout design of the power-gating router implemented in Sect. 2.4.2. The run-time power gating is applied to VC buffer, Output latch, VCMUX, and CBMUX. We used the 106 customized standard cells based on a low-power version of a commercial 65 nm standard cell library.[1] Temperature and core voltage are set to 25°C and 1.20 V, respectively. These leakage parameters are fed to the full system CMP simulator, in order to evaluate the run-time leakage power of the routers when the application programs run on them.

Table 2.5 Leakage power of router components [uW]

Module	Count	Total leakage power
4-flit VC buffer	20	189.07
1-flit output latch	5	16.71
5-to-1 CBMUX	5	11.41
4-to-1 VCMUX	5	13.45
Others	1	38.36
Total		269

[1] We selected a low-power CMOS process whose leakage power is quite small, since our final goal is to develop ultra low leakage on-chip networks.

To clearly demonstrate the leakage power reduction of each power domain type, the proposed fine-grained power gating is gradually applied to the router as the following three steps.

- **Level 1:** Only VC buffers are power gated.
- **Level 2:** VC buffers, VCMUXes, and CBMUXes are power gated.
- **Level 3:** VC buffers, VCMUXes, CBMUXes, and Output latches are power gated.

Figure 2.12a shows an average leakage power of the router when Level-1 power gating, which covers only VC buffers, is used. The wakeup latency of all power domains is set to three cycles assuming an operating frequency of 1 GHz. In this graph, 100% indicates the leakage power of the router without power gating (i.e., 269 uW).

The original Look-ahead method shows the smallest leakage power in these methods, while it cannot hide the first-hop wakeup latency and degrades the application performance, as shown in Sect. 2.6.2. Look-ahead with ABW method consumes the largest leakage power since an active buffer window (i.e., two flits) of each VC buffer is always activated, although it achieves the best performance. Look-ahead with ever-on method consumes a little bit more leakage power compared to the original Look-ahead method, since it has some ever-on power domains to mitigate the first-hop wakeup latency. Fortunately, the leakage power of these ever-on domains is not crucial, since they are limited to VC buffers of VC0 and VC2 in input physical channels directly connected from processor cores. As a result, Look-ahead with ever-on method is the best choice to balance the tradeoff between the performance and leakage power. In Level-1 power gating, it reduces the average router leakage power by 64.6%.

Figure 2.12b shows an average leakage power of the router when Level-2 power gating, which covers VC buffers, VCMUXes, and CBMUXes, is applied. The same tendency as Level 1 can be seen in Level-2 power gating. Look-ahead with ever-on method reduces the average router leakage power by 72.7%.

Figure 2.12c shows an average leakage power of the router when our fine-grained power gating is fully applied. In Level-3 power gating, Look-ahead with ever-on method reduces the average router leakage power by 77.7%.

As a result, our fine-grained power gating with Look-ahead ever-on method reduces the leakage power by 77.7% even when application programs are running, at the expense of 4.0% performance overhead when we assume a 1 GHz operation. The area overhead is 4.3% as estimated in Sect. 2.4.2. Thus, our fine-grained power-gating router presents favorable trade-offs between these modest overheads and the significant leakage power saving.

2.7 Summary

The content of this chapter can be summarized as follows:

- An on-chip router architecture is illustrated and its power consumption is analyzed.

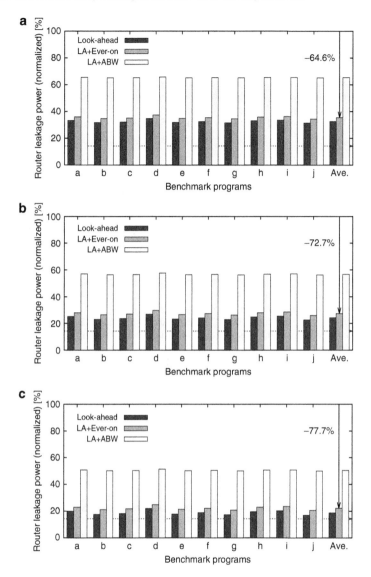

Fig. 2.12 Average leakage power of an on-chip router with different wakeup methods (100% indicates the leakage power without power gating). (**a**) Power domains: VC buffer (**b**) Power domains: VC buffer, VCMUX, and CBMUX (**c**) Power domains: VC buffer, VCMUX, CBMUX, and Output latch

- Low-power techniques especially power gating for NoCs are surveyed.
- A fine-grained power gating router with 35 micro-power domains using a commercial 65 nm process is introduced.
- The fine-grained power gating router is evaluated in terms of area, wakeup latency, application performance, and leakage power reduction.

As future work, we are planning to apply the fine-grained power gating to the network interfaces. We are considering the use of various information from coherence protocol or operating system to guide the run-time power gating decisions.

Acknowledgements This research was performed by the authors for STARC as part of the Japanese Ministry of Economy, Trade and Industry sponsored "Next-Generation Circuit Architecture Technical Development" program. The authors thank VLSI Design and Education Center (VDEC) and Japan Science and Technology Agency (JST) CREST for their support.

References

1. Agarwal, N., Peh, L.S., Jha, N.: Garnet: A Detailed Interconnection Network Model Inside a Full-System Simulation Framework. Tech. Rep. CE-P08-001, Princeton University (2008)
2. Banerjee, A., Mullins, R., Moore, S.: A Power and Energy Exploration of Network-on-Chip Architectures. In: Proceedings of the International Symposium on Networks-on-Chip (NOCS'07), pp. 163–172 (2007)
3. Beigne, E., Clermidy, F., Lhermet, H., Miermont, S., Thonnart, Y., Tran, X.T., Valentian, A., Varreau, D., Vivet, P., Popon, X., Lebreton, H.: An Asynchronous Power Aware and Adaptive NoC Based Circuit. IEEE Journal of Solid-State Circuits **44**(4), 1167–1177 (2009)
4. Chen, X., Peh, L.S.: Leakage Power Modeling and Optimization in Interconnection Networks. In: Proceedings of the International Symposium on Low Power Electronics and Design (ISLPED'03), pp. 90–95 (2003)
5. Dally, W.J.: Virtual-Channel Flow Control. IEEE Transactions on Parallel and Distributed Systems **3**(2), 194–205 (1992)
6. Galles, M.: Spider: A High Speed Network Interconnect. IEEE Micro **17**(1), 34–39 (1997)
7. Hu, Z., Buyuktosunoglu, A., Srinivasan, V., Zyuban, V., Jacobson, H., Bose, P.: Microarchitectural Techniques for Power Gating of Execution Units. In: Proceedings of the International Symposium on Low Power Electronics and Design (ISLPED'04), pp. 32–37 (2004)
8. IEEE Standard for Design and Verification of Low Power Integrated Circuits. IEEE Computer Society (2009)
9. Ikebuchi, D., Seki, N., Kojima, Y., Kamata, M., Zhao, L., Amano, H., Shirai, T., Koyama, S., Hashida, T., Umahashi, Y., Masuda, H., Usami, K., Takeda, S., Nakamura, H., Namiki, M., Kondo, M.: Geyser-1: A MIPS R3000 CPU Core with Fine Grain Runtime Power Gating. In: Proceedings of the IEEE Asian Solid-State Circuits Conference (A-SSCC'09) (2009)
10. Ishikawa, M., Kamei, T., Kondo, Y., Yamaoka, M., Shimazaki, Y., Ozawa, M., Tamaki, S., Furuyama, M., Hoshi, T., Arakawa, F., Nishii, O., Hirose, K., Yoshioka, S., Hattori, T.: A 4500 MIPS/W, 86 µA Resume-Standby, 11 µA Ultra-Standby Application Processor for 3G Cellular Phones. IEICE Transactions on Electronics **E88-C**(4), 528–535 (2005)
11. Kanno, Y., et al.: Hierarchical Power Distribution with 20 Power Domains in 90-nm Low-Power Multi-CPU Processor. In: Proceedings of the International Solid-State Circuits Conference (ISSCC'06), pp. 2200–2209 (2006)
12. Kim, C., Burger, D., Keckler, S.W.: An Adaptive, Non-Uniform Cache Structure for Wire-Delay Dominated On-Chip Caches. In: Proceedings of the International Conference on Architectural Support for Programming Languages and Operating Systems (ASPLOS'02), pp. 211–222 (2002)
13. Magnusson, P.S., et al.: Simics: A Full System Simulation Platform. IEEE Computer **35**(2), 50–58 (2002)
14. Martin, M.M.K., Hill, M.D., Wood, D.A.: Token Coherence: Decoupling Performance and Correctness. In: Proceedings of the International Symposium on Computer Architecture (ISCA'03), pp. 182–193 (2003)

15. Martin, M.M.K., Sorin, D.J., Beckmann, B.M., Marty, M.R., Xu, M., Alameldeen, A.R., Moore, K.E., Hill, M.D., Wood, D.A.: Multifacet General Execution-driven Multiprocessor Simulator (GEMS) Toolset. ACM SIGARCH Computer Architecture News (CAN'05) 33(4), 92–99 (2005)
16. Matsutani, H., Koibuchi, M., Wang, D., Amano, H.: Adding Slow-Silent Virtual Channels for Low-Power On-Chip Networks. In: Proceedings of the International Symposium on Networks-on-Chip (NOCS'08), pp. 23–32 (2008)
17. Matsutani, H., Koibuchi, M., Wang, D., Amano, H.: Run-Time Power Gating of On-Chip Routers Using Look-Ahead Routing. In: Proceedings of the Asia and South Pacific Design Automation Conference (ASP-DAC'08), pp. 55–60 (2008)
18. Matsutani, H., Koibuchi, M., Amano, H., Yoshinaga, T.: Prediction Router: Yet Another Low Latency On-Chip Router Architecture. In: Proceedings of the International Symposium on High-Performance Computer Architecture (HPCA'09), pp. 367–378 (2009)
19. Mullins, R., West, A., Moore, S.: Low-Latency Virtual-Channel Routers for On-Chip Networks. In: Proceedings of the International Symposium on Computer Architecture (ISCA'04), pp. 188–197 (2004)
20. Seki, N., Zhao, L., Kei, J., Ikebuchi, D., Kojima, Y., Hasegawa, Y., Amano, H., Kashima, T., Takeda, S., Shirai, T., Nakata, M., Usami, K., Sunata, T., Kanai, J., Namiki, M., Kondo, M., Nakamura, H.: A Fine-Grain Dynamic Sleep Control Scheme in MIPS R3000. In: Proceedings of the International Conference on Computer Design (ICCD'08), pp. 612–617 (2008)
21. Shang, L., Peh, L.S., Jha, N.K.: Dynamic Voltage Scaling with Links for Power Optimization of Interconnection Networks. In: Proceedings of the International Symposium on High-Performance Computer Architecture (HPCA'03), pp. 79–90 (2003)
22. Soteriou, V., Peh, L.S.: Exploring the Design Space of Self-Regulating Power-Aware On/Off Interconnection Networks. IEEE Transactions on Parallel and Distributed Systems 18(3), 393–408 (2007)
23. Stine, J.M., Carter, N.P.: Comparing Adaptive Routing and Dynamic Voltage Scaling for Link Power Reduction. IEEE Computer Architecture Letters 3(1), 14–17 (2004)
24. Usami, K., Ohkubo, N.: A Design Approach for Fine-grained Run-Time Power Gating using Locally Extracted Sleep Signals. In: Proceedings of the International Conference on Computer Design (ICCD'06) (2006)
25. Vangal, S.R., Howard, J., Ruhl, G., Dighe, S., Wilson, H., Tschanz, J., Finan, D., Singh, A., Jacob, T., Jain, S., Erraguntla, V., Roberts, C., Hoskote, Y., Borkar, N., Borkar, S.: An 80-Tile Sub-100-W TeraFLOPS Processor in 65-nm CMOS. IEEE Journal of Solid-State Circuits 43(1), 29–41 (2008)
26. Woo, S.C., Ohara, M., Torrie, E., Singh, J.P., Gupta, A.: SPLASH-2 Programs: Characterization and Methodological Considerations. In: Proceedings of the International Symposium on Computer Architecture (ISCA'95), pp. 24–36 (1995)

Chapter 3
Adaptive Voltage Control for Energy-Efficient NoC Links

Paul Ampadu, Bo Fu, David Wolpert, and Qiaoyan Yu

Abstract As we enter the many-core integration era driven by advances in multiprocessor system-on-chip innovations, interconnect emerges as the bottleneck in achieving energy efficiency in systems-on-chip. This chapter surveys the state-of-the-art in energy-efficient communication link design for NoCs. After reviewing techniques at the datalink and physical abstraction layers, we introduce a lookahead-based transition-aware adaptive voltage control method for achieving improved energy-efficiency at moderate cost in performance and reliability. Limitations of this method are evaluated and future prospects in energy-efficient link design are projected.

3.1 Introduction

While global interconnect dimensions have scaled much more slowly than transistor dimensions, the number of interconnect-related issues has increased with technology scaling because of increased operating speeds, increased energy densities resulting in larger on-chip thermal gradients, and reduced supply voltages increasing the impact of noise sources [1]. Coupling between data lines results in a huge data dependency on speed, power, and energy [2]. Long links increase the impact of intra-die variation, which is generally small at the local level but can result in large delay and power performance differences across a chip. In high frequency nanoscale systems, interconnect performance can become a function of inductance as well as capacitance and resistance, increasing the complexity of modeling and verification [3]. In addition, as technology scales, interconnect power is becoming an increasingly large component of total chip power dissipation, making power- and energy-efficiency one of the most important design metrics in nanoscale interconnect design [4].

P. Ampadu (✉)
University of Rochester, Rochester, NY 14627, USA
e-mail: ampadu@ece.rochester.edu

C. Silvano et al. (eds.), *Low Power Networks-on-Chip*,
DOI 10.1007/978-1-4419-6911-8_3, © Springer Science+Business Media, LLC 2011

A large amount of research has focused on finding ways to improve interconnect energy performance. The network infrastructure has a large cost – up to 36% of the power dissipation in each networked tile [4]. One of the most common techniques involves controlling the link swing voltage, either through static optimization or adaptive voltage schemes. Other techniques include data pattern optimization, repeater insertion, and novel transmission methods such as current-mode signaling or pulse transmission. In this chapter, we will examine each of these techniques, providing a special emphasis on a recently proposed adaptive voltage scheme that takes advantage of advanced knowledge of data patterns to reduce power dissipation in on-chip interconnect links.

3.2 Methods for Energy-Efficient On-Chip Links

The techniques used to improve energy-efficiency in on-chip links span multiple layers of the NoC design space. In this section, we provide background on commonly used techniques at both the datalink and physical layers and describe a set of metrics we will use throughout this chapter.

3.2.1 Metrics for Energy Efficiency

Before discussing specific techniques that have been proposed to improve energy efficiency in on-chip interconnect, let us briefly explain the metrics used in this chapter. Naturally, power dissipation and energy dissipation are commonly used metrics [5–7], where energy E is the product of power dissipation P and time t. The energy consumption of a wire E_{wire} is the product of activity factor α, effective wire capacitance C_{eff}, supply voltage V_{DD}, and link swing voltage V_{swing} [8] ($E_{wire} = \alpha \cdot C_{eff} \cdot V_{DD} \cdot V_{swing}$). To compare the energies of different methods, peak energies are measured by applying worst case switching patterns to the link [9]. Alternatively, average case energies can be obtained by applying every input pattern to the link and averaging the dissipated energies [10] (assuming the input patterns are normally distributed).

When referring to power and energy, we must specify which system components we are including. For example, Zhang et al. [11] report the total energy consumed by the entire transmitter and receiver, while Benini et al. [12] report the minimum power dissipation per bus transition for only their low power encoder and decoder. If on-chip noise sources corrupt a transmitted message, it can be dropped, corrected or a retransmission may be requested. In these cases, energy per transmission may be a misleading metric; a better metric is energy per useful flit, which includes the ratio of the total transmission energy and successfully transferred flits [13]. Worm et al. [14] use the energy per transmitted word to provide a complete overview of the system energy dissipation, including the codec energy, the energy for transmitting a message, the energy for transmitting redundancy bits created by the applied code, control logic for the encoder/decoder, system synchronization overhead, voltage converters on the link, and any retransmission energy.

We must also be careful to indicate the number of wires in a link we are describing, as well as the number of pipeline stages in the link; in [15], the total energy of an n-cycle bus that is i wires wide is reported. In some applications, energy consumption is an inappropriate metric because large energy improvements can be achieved at the expense of large delays. To compare the energy in systems where more stringent delay requirements must be met, the metric of choice is energy-delay product [11, 16].

3.2.2   Datalink Layer Techniques

Datalink layer techniques describe improvements to NoC switches that modify the transmitted data using some form of coding. The applied coding results in an increase in area overhead and adds some latency to the transmitter and receiver pipelines. Coding incurs some power and energy overhead as well, however the energy-efficient codes described in this section provide a net energy benefit by saving more energy on the link than they consume in the codec.

3.2.2.1   Bus Invert and Extended Bus Invert Coding

One of the more straight-forward methods for reducing link energy is bus-invert coding [17]. In bus-invert coding, the encoder determines the percentage of bus wires that must be flipped to send the next flit. If this percentage is greater than 50%, the input data are inverted and an additional wire is used to communicate that the data have been inverted, as shown in Fig. 3.1. This technique reduces peak power dissipation by nearly 50% (depending on the width of the link) and average power dissipation by nearly 25%. The data inversion technique has also been extended into partial bus-invert coding, an application-specific technique that groups wires with highly correlated switching patterns and high transition probabilities to further improve the efficiency of bus invert coding [18]. This technique results in an improvement of up to 72% beyond what is achievable by applying bus-invert coding to the entire bus.

3.2.2.2   Frequent Value Coding

Instead of using a control signal to determine if the data should be inverted (as in bus-invert coding), frequent value (FV) coding creates a small table of statically or dynamically determined common bit patterns, and uses a small look-up table to create one-hot encoding alternatives (setting one wire to "1" and the rest to "0")

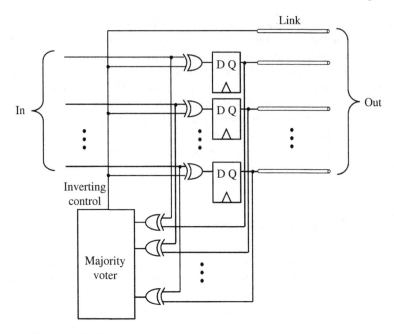

Fig. 3.1 Diagram of bus-invert coding encoder

for these high-transition patterns [19]. By XORing the one-hot alternatives with the previous data transmissions, FV coding is able to reduce switching between two and four times as effectively as bus-invert coding.

3.2.2.3 Crosstalk Avoidance Coding

As coupling capacitances increase with technology scaling, the additional power consumed by coupling becomes a large portion of the total power dissipation. Instead of reducing power dissipation by minimizing the number of transitions, crosstalk avoidance coding has been proposed to reduce the power consumed by coupling capacitance. In crosstalk avoidance codes (CACs) [2,20], the input data are mapped to a CAC codeword. The CAC codewords are designed to avoid switching patterns with large coupling capacitance, reducing total power consumption by up to 40% [20]. Detection of crosstalk patterns may also be used to set the cycle period for average case crosstalk latency, improving performance up to 31% [21]; when worst-case crosstalk patterns are detected, the transmissions are split over multiple cycles.

3.2.2.4 Asymptotic Zero-Transition Coding

In the gray code, only one bit differs between consecutive binary numbers. Based on the observation that input patterns on address busses are consecutive,

Benini et al. [12] use the gray code for consecutive addresses and add an additional redundant line to indicate non-consecutive addresses. In addition to this low-switching encoding scheme, the receiver can be designed to automatically calculate the next address without switching the address busses in the majority of cases, greatly reducing switching energy (in the best case, switching is reduced to an average of less than one transition per flit).

3.2.3 Physical Layer Techniques

Physical layer techniques describe performance improvements in the circuits and wires, including changes to the sizing and spacing of wires, changes to the applied voltages on the link drivers and repeaters, and adjustments to driver, repeater, and receiver designs.

3.2.3.1 Low-Swing Signaling

As mentioned in Sect. 3.2.1, energy dissipation is directly proportional to both the link swing voltage V_{swing} and the supply voltage V_{DD}. Reducing the supply voltage on the link also reduces V_{swing}, quadratically decreasing dynamic energy dissipation E_{dyn}, as shown in (3.1):

$$E_{dyn} = \int_0^\infty i(t)V_{DD}\mathrm{d}t = V_{DD}\int_0^\infty C_L\frac{\mathrm{d}v_{out}}{\mathrm{d}t}\mathrm{d}t = C_L V_{DD}\int_0^{V_{swing}}\mathrm{d}v_{out}$$
$$= C_L V_{DD} V_{swing} \tag{3.1}$$

where E_{dyn} is dynamic energy dissipation, $i(t)$ is the current drawn from the supply voltage, C_L is the load capacitance, and v_{out} is the driver output voltage. A major drawback of these low voltage systems is increased transition latency (latency is loosely approximated by $C_{eff}V_{swing}/(2I_{on})$, where I_{on} is the device on-current).

Low-swing drivers and receivers have been extensively applied to interconnect links [11, 16, 22], achieving energy improvements in excess of 5× for a given frequency target [11]. Rather than reducing the supply voltage to swing between 0 V and V_{swing}, an alternative low-swing technique centers the link voltage at the switching threshold ($\sim V_{DD}/2$) and uses specially designed receivers that detect small changes in voltage swing [16]. As shown in Fig. 3.2, low-swing drivers can be as simple as an inverter with a reduced PMOSFET source voltage; low-swing receivers are slightly more complex, requiring level converters to restore the link voltage to that of the receiver flip-flop. Without this level converter, the flip-flop input stage will only be partially shut-off, resulting in large short-circuit currents [11, 22]. This low-impedance termination results in energy improvements of up to 56% compared with a conventional interconnect with repeater insertion, while also improving latency by up to 21% [16].

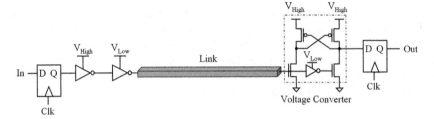

Fig. 3.2 Diagram of low-swing link transmitter and receiver

3.2.3.2   Differential Signaling

In differential signaling, two wires are used to transmit one data bit. The receiver recovers the input data by sensing the voltage or current difference between the two wires [23]. One major benefit of the two-wire approach is common-mode noise reduction. This reduction in noise sources can allow lower voltage swings to be used, resulting in energy savings in excess of 83% and power-delay product savings of 81% despite the doubling of the number of link wires [11]. Also, differential drivers tend to be much smaller than standard drivers, which results in a reduction in power consumption (a 66% improvement in transmitter power is reported in [24]). Additional improvements to differential signaling include the use of a capacitive pre-emphasis circuit [24] to drive the link rather than a traditional inverter; this method has been shown to improve data rates by 66% compared to a prior differential signaling approach while also achieving a 4× reduction in link power.

3.2.3.3   Repeater Insertion

Repeater insertion is commonly used to minimize interconnect delay by restoring attenuated signals. In general, repeaters that are optimized for delay can be quite large and dissipate a significant amount of power. To reduce the power consumption of repeaters, smaller and fewer repeaters can be used for noncritical interconnects with a small delay penalty but a large power saving [25]. Alternatively, the repeaters can be designed specifically to meet a target power budget [26]. Skewed repeaters can also be used to reduce crosstalk-coupling power, resulting in link energy improvements of up to 18% compared to conventional repeater insertion [15]. Weeasekera et al. have recently proposed a dynamic repeater insertion technique where the repeater strength is adjusted depending on the calculated switching capacitance of each transition, resulting in average energy savings of 20–25% compared to conventional repeater insertion [10]. As leakage power dissipation becomes a larger component of the total power dissipation, reducing repeater size and the number of repeaters in noncritical interconnects can also be used to reduce the leakage power.

3.2.3.4   Dual-Voltage Buffers

There are a few varieties of dual-voltage buffers, but most can be grouped into either dual-threshold voltage (dual-V_T) designs [5, 27] or dual-supply voltage (dual-V_{DD}) designs [28–30]. Optimal selection of devices from two thresholds has resulted in power savings of up to 40% compared to a single-V_T approach [5]. Setting adjacent wires to have different threshold voltages also reduces the delay penalty of the worst-case coupling capacitance, allowing smaller device sizing and resulting in energy improvements of up to 31% [27].

Dual-V_{DD} designs have been used in FPGAs to avoid the need for level converters by identifying entire paths that can be operated at the low voltage, which has resulted in interconnect power savings of ~53% compared to a single-V_{DD} implementation [28]. More fine-grained implementations of dual-V_{DD} approaches apply both supply voltages to each wire driver, switching to the high voltage for a short period of time when transitions are detected and switching back to the low voltage before the entire link can be charged. This technique has resulted in energy improvements of up to 17% [29]. The fine-grained dual-voltage technique has also been combined with pulsed transmission (described in the following subsection), to reduce link power by up to 50% compared to non-pulsed, single-V_{DD} approaches [30].

3.2.3.5   Pulsed Transmission

Voltage-controlled long global interconnect imposes very heavy loads on the drivers, requiring repeaters to reduce the propagation delay at the cost of increasing area and power consumption. By carefully considering the link impedance, pulsed transmission techniques can be used, which replace on-chip links and repeaters with low latency, low energy interconnect, modulating the link transmission using short current-pulses to achieve higher frequencies [31–33] than possible by fully charging and discharging the link. The pulsed transmission system is shown in Fig. 3.3 and shows how the digital input is converted to a pulse for transmission on the link.

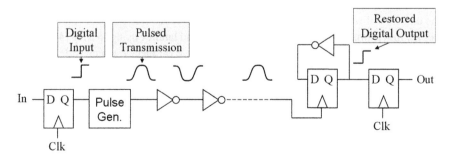

Fig. 3.3 Diagram of pulsed transmission system

The pulse triggers the toggle flip-flop at the receiver, which then outputs the appropriate digital value. This pulsed wave interconnect only partially charges the link, saving energy by up to 50% compared to the repeater insertion method.

3.2.3.6 Current Mode Signaling

Current mode signaling involves transmitting information using different current levels rather than different voltage levels. These techniques achieve delay improvements of up to 50% over voltage-mode approaches while matching power consumption [34]. At high data rates, current mode signaling can also result in significant power savings [34] ($\sim$50% power savings at $700\,\mathrm{Mb\,s^{-1}}$ in a $0.13\,\mu m$ technology). At low data rates, these power savings are reduced by the static power dissipation in the current sensing receiver [35].

3.2.3.7 Globally Asynchronous Locally Synchronous (GALS) Signaling

Clock distribution and synchronization of large chip multiprocessors become increasingly challenging as the number of cores scales, resulting in large energy overheads and large intra-die clock skews that limit performance. One technique of addressing these issues is to allow each core to operate on its own clock while connecting the cores using asynchronous handshaking protocols. This technique is referred to as globally asynchronous, locally synchronous (GALS) communication and is increasingly needed to manage the multitude of processing cores and memory elements that each have individual performance requirements [36]. The benefits of GALS approaches include increased modularity and scalability, per-core performance optimization (rather than using a global worst-case clock frequency), reduction in clock power, and reduction in clock skew [37]. GALS systems have been designed using a variety of topologies, such as a mesh-based network-on-chip [38] or a 16-port central crossbar [39].

3.2.3.8 Quasi-Resonant Interconnect

The addition of on-chip spiral inductors to long-distance interconnect can improve average link power consumption by up to 91.1% and latency by up to 37% [40]. This is achieved by operating the link at a fixed resonance frequency, which is dependent on the link impedance. The impressive power savings are achieved by resonating the energy between the electromagnetic fields, which reduces the portion of energy that is dissipated as heat. The major drawback of this approach is the area requirement for the inductors; in a 50 nm technology, the power and latency benefits of the spiral inductor come at an area cost of $\sim$10$\times$ that of repeater insertion for a 1 mm link.

3.2.3.9 Adaptive Link Voltage Scaling

Adaptive link voltage schemes have been used to achieve average-case performance rather than worst-case performance, adjusting the voltage and clock frequency to meet a wide range of performance targets [22]. For example, the link voltage may be adjusted based on the application; a history-based voltage scaling approach resulted in average energy savings of 4.6× compared to a fixed approach [41]. In addition, the ability to adjust the supply voltage can be used to provide variation tolerance; in [7], the supply voltage is adjusted based on detected process variations to ensure that each link can operate at the desired frequency. In [6], this variation tolerance is combined with a low-swing approach, and the optimal number of voltage steps are calculated, resulting in a net power improvement of 43% over a fixed voltage approach for a 64-bit link.

3.2.4 Other Methods

In Sects. 3.2.2 and 3.2.3, we discussed a variety of data link layer and physical layer approaches. In this subsection, we address additional alternatives that combine both datalink layer and physical layer approaches.

3.2.4.1 Integrating Double Sampling with Adaptive Voltage Scaling

Variations in interconnect parameters and operating environments require large safety margins to ensure that worst-case design targets are achieved. These additional safety margins reduce the power efficiency of adaptive voltage scaling. In [42], a more aggressive voltage scaling technique is proposed to operate at average-case conditions. In this method, doubling sampling latches are used to detect and correct timing errors in the links, and voltage scaling is controlled by the error recovery rates provided by the latches. By eliminating the additional safety margins for worst-case design, this method can further reduce power consumption by 17% over conventional adaptive voltage scaling even when worst-case variations are considered. During average-case variations, energy savings of up to 45% have been reported.

3.2.4.2 Combining Error Control Coding with Adaptive Voltage Scaling

Although reduced link swing voltages reduce link energy consumption, decreases in link swing voltage also reduce link reliability. One approach to solve this problem is to combine error control methods with low link swing voltage systems [13, 14, 43], as shown in Fig. 3.4. In these methods, the link swing voltage can be selected to

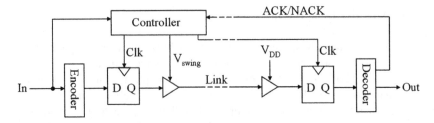

Fig. 3.4 Block diagram of integrated error control coding and adaptive voltage scaling system

meet a target system reliability requirement. More powerful error control coding schemes can achieve lower link swing voltages, increasing the potential power and energy savings [43].

3.3 Lookahead-Based Transition-Aware Link Voltage Control

An adaptive voltage scheme that uses advanced knowledge of data patterns to reduce power dissipation in on-chip interconnect links has been proposed [44]. In most NoC links, the data to be transmitted are known ahead of time, in some cases multiple cycles in advance. This advanced knowledge of incoming data patterns offers an opportunity to look ahead at the data patterns and precondition the link to improve interconnect delay, power, and/or energy performance.

This section describes a system that uses lookahead information to detect incoming transitions on each wire and select between two driver supply voltages, V_{Low} and V_{High}. A simple transition detection circuit uses incoming data patterns to adjust the driver voltage. In the lookahead implementation, a reduced link voltage is used in the general case, and when a rising transition is detected, the driver supply voltage is raised for a small portion of the clock cycle. This high-voltage pulse improves the circuit's rising transition delay, but the high voltage is only applied for a fraction of the clock cycle and is not intended to charge the entire link capacitance to V_{High}. The supply voltage boost allows this approach to improve the delay performance of a low-swing interconnect system, while maintaining the energy benefits of the reduced swing voltage. While many previous dual-voltage schemes adjust the voltage on the entire bus (with notable exceptions [29]), this approach provides a fine-grain control of the voltage on each wire in the link. A block diagram of the lookahead transition-aware system is shown in Fig. 3.5. The current wire state (at cycle t) and the state of the data in the next cycle (cycle $t+1$) are inputs to the transition detection unit, which is used to adaptively tune the supply voltage of the transmitter driver. The transition detection information is used to adjust the voltages on rising transitions.

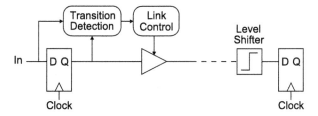

Fig. 3.5 Block diagram of transition-aware, lookahead-based adaptive system

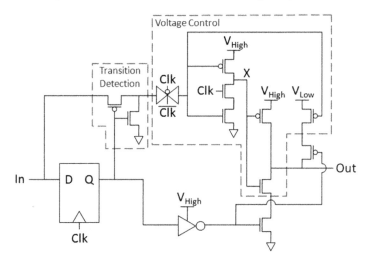

Fig. 3.6 Lookahead transmitter schematic

3.3.1 Lookahead Transmitter Design

The transition detection and voltage control circuits are shown in Fig. 3.6. The transition detection circuit is used to detect a rising transition (a "0" at node Q, the output of the register, and a "1" at node In, the input to the register). In the transition detection circuit, the PMOS device is enabled when $Q = 0$, passing In to the transmission gate. In all other cases, the NMOS device is enabled and the transition detection output is pulled down to ground; thus, the only case when the transition detection output is "1" is when $Q = 0$ and $In = 1$.

The duration of V_{High} can be adjusted depending on performance requirements. In this implementation, V_{High} is applied for half of the clock cycle. The transmission gate latches the output of the transition detector when $Clk = 0$. This is needed to prevent a change in the value of In from affecting the voltage boost during the rising transition. The latched output of the transition detection circuit is then passed into a low-skewed inverter with a clocked NMOS device to prevent the inverter from triggering until the beginning of the clock cycle (the output of the transition

Fig. 3.7 Lookahead
transmitter timing waveforms

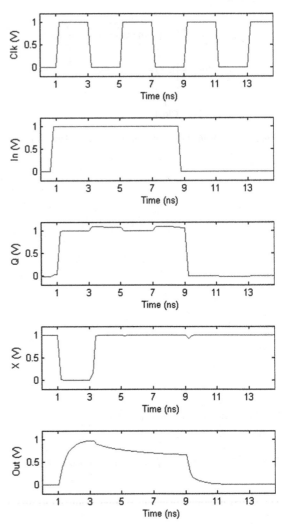

detection circuit reaches the input of the skewed inverter at the falling edge of the previous cycle). The inverter is skewed to improve the speed at which node X is pulled down.

The timing diagrams in Fig. 3.7 show the behavior of the lookahead transmitter in more detail. At $t = 0.9$ ns, the input switches from low to high, activating the transition detection circuit. On the next rising edge of the clock ($t = 1$ ns), the transition status is latched by the transmission gate. Node X is driven low, and V_{High} is applied to the final buffer stage until the falling edge of the clock ($t = 3$ ns). During this time, the buffer output *Out* is charged to V_{High}, and the link output decays to V_{Low} once the precharge node X returns to logic high.

The transition between V_{High} and V_{Low} results in the pattern shown in the *Out* waveform in Fig. 3.7. V_{High} is used to improve the rising delay of the output;

however, the link capacitance is not necessarily fully charged to V_{High}. Applying the high voltage for only a portion of the cycle reduces the link swing voltage, resulting in significant energy savings compared to the traditional approach (nominal voltage with a two-inverter buffer). In the lookahead scheme, the falling edge constraint, $T_{1\rightarrow0}$, is shown in (3.2).

$$T_{1\rightarrow0} = T_{Clk\rightarrow Q} + T_{int\_inv,0\rightarrow1} + T_{driver,1\rightarrow0} + T_{trans,1\rightarrow0} + T_{lvlshft,1\rightarrow0}$$
$$+ T_{D\rightarrow Clk} \tag{3.2}$$

$T_{Clk\rightarrow Q}$ is the transmitter flip-flop clock-to-output latency; $T_{int\_inv,0\rightarrow1}$ is the rising delay through the internal inverter in the transmitter; $T_{driver,1\rightarrow0}$ is the falling delay of the link driver; $T_{trans,1\rightarrow0}$ is the signal propagation latency on the wire; $T_{lvlshft,1\rightarrow0}$ is the level shifter falling latency; and $T_{D\rightarrow Clk}$ is the hold time requirement of the receiver flip-flop.

For the rising transition constraint in the lookahead scheme, $T_{0\rightarrow1}$, the flip-flop is bypassed, and that portion of the constraint is replaced by the latency of the clocked inverter controlling node X in Fig. 3.6. The rising transition latency is thus the time it takes for node X to stabilize, plus the delay through the driver pull-up inverter $T_{driver,0\rightarrow1}$; the rising edge constraint in the lookahead scheme $T_{0\rightarrow1}$ is shown in (3.3).

$$T_{0\rightarrow1} = T_{Clk\rightarrow X} + T_{driver,0\rightarrow1} + T_{trans,0\rightarrow1} + T_{lvlshft,0\rightarrow1} + T_{D\rightarrow Clk} \tag{3.3}$$

A waveform highlighting the improvement in rising transition delay is shown in Fig. 3.8; the lookahead system's link voltage reaches 0.5 V over 70 ps earlier than a single voltage scheme using $V_{High} = 1$ V (referred to as the traditional approach in Fig. 3.8), a 44% improvement. Figure 3.8 was generated in Cadence Spectre using a 45 nm predictive technology model (PTM) [45].

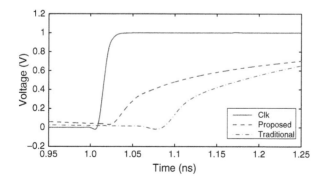

Fig. 3.8 Waveform comparison of rising transition between lookahead scheme and single voltage scheme using $V_{High} = 1$ V

The lookahead method has the setup time constraint shown in (3.4):

$$T_{D \to Clk} \geq max(T_{D \to Clk}, T_{tran\_det} + T_{TG} + T_{clk\_inv}) \tag{3.4}$$

where T_{tran_det} is the delay through the transition detection circuit, T_{TG} is the transmission gate latency, and T_{clk_inv} is the setup time of the clocked inverter controlling node X. The addition of T_{tran_det} and T_{TG} increase the setup time constraint by 30 ps over the standard flip-flop setup time.

3.3.2 HI/LO Voltage Selection

One of the most important design decisions in a multivoltage system is the selection of the operating voltages. The choice of voltage impacts nearly every important design metric, including speed, power, energy, area (i.e., sizing requirements to meet a given frequency target), and reliability.

This approach is intended to improve the power and energy performance of an NoC link, and the energy dissipations of the approach are shown in Fig. 3.9 for a variety of operating voltages. All the presented circuit simulations were generated in Cadence Spectre using a 45 nm PTM [45]. The x-axis in Fig. 3.9 represents different values of V_{High}, while the y-axis indicates values of V_{Low} between $0.6*V_{High}$ and $0.9*V_{High}$. For example, the point where x = 0.9 and y = 0.8 refers to the point where $V_{High} = 0.9$ V and $V_{Low} = 0.8*0.9$ V = 720 mV. Values of $V_{Low} < 0.6*V_{High}$ are not shown because the receiver flip-flop is assumed to be operating at V_{High}, and input voltages lower than $0.6*V_{High}$ may be insufficient to switch the flip-flop state.

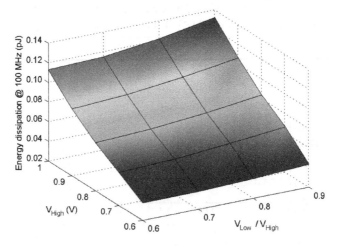

Fig. 3.9 The impact of voltage selection on the energy dissipation of the lookahead approach at 100 MHz

Table 3.1 Power dissipation of lookahead approach over a range of operating voltages and frequency targets

		Target frequency (GHz)						
V_{High}	V_{Low}	2.0	1.75	1.5	1.25	1.0	0.75	0.5
1.0 V	0.6 V	–	1.78 mW	1.65 mW	1.52 mW	1.34 mW	1.12 mw	0.79 mW
0.9 V	0.6 V	–	1.41 mW	1.32 mW	1.22 mW	1.09 mW	0.92 mW	0.66 mW
0.8 V	0.6 V	–	–	0.97 mW	0.91 mW	0.83 mW	0.72 mW	0.54 mW
0.7 V	0.6 V	–	–	–	–	–	0.51 mW	0.41 mW

One obvious conclusion to draw from Fig. 3.9 is that lower voltages provide lower energy dissipation, but there is another important point to draw from the data. For a given V_{High}, reducing the value of V_{Low} from $0.9*V_{High}$ all the way to $0.6*V_{High}$ results in a smaller energy savings than reducing V_{High} by 100 mV. For example, the energy consumed by the system with $V_{High} = 1$ V and $V_{Low} = 0.6$ V is larger than the energy consumed by the system with $V_{High} = 0.9$ V and $V_{Low} = 0.9*V_{High} = 810$ mV. In that case, reducing V_{Low} by over 200 mV has less impact on energy dissipation than reducing V_{High} by 100 mV.

Reducing V_{Low} has very little effect on the rising edge delay because V_{High} is used to boost the link voltage. Reducing V_{Low} can also improve the falling edge delay because of the lower amount of stored charge to dissipate. This information, combined with the data from Fig. 3.9, can be used to provide an easy guideline for optimizing the voltages in the lookahead system. When using this approach to optimize energy dissipation, one should first find the smallest V_{High} that satisfies the given frequency requirement, and then reduce V_{Low} to the lowest point at which the receiver flip-flop can be controlled (with an appropriate design margin).

To illustrate this method, Table 3.1 provides a number of target frequencies and the energy dissipations at four values of V_{High}. In this study, V_{Low} was fixed to 0.6 V, which is one half of the nominal voltage with an additional 10% voltage noise precaution. For a target frequency of 1.5 GHz, a V_{High} of 0.8 V is the most efficient choice, providing an energy savings of $\sim$40% beyond that achieved using the lookahead system at nominal V_{High}. As the frequency target is reduced to 0.75 GHz, $V_{High} = 0.7$ V is the most efficient choice, providing an energy savings of over 53%.

3.3.3 Performance Evaluation

The link model used in all of the presented simulations is a global wire broken into RC segments, with each segment representing a link length of 1 mm. The global wire parameters and parasitic capacitance values are shown in Table 3.2 with a cross-section shown in Fig. 3.10. Resistance and parasitic capacitance values (R, C_g, and C_C) were calculated [45] using a 45 nm global link interconnect model [46]. No inductances are included in the parasitic model because of the large signal transition time, t_r (the line lengths in question are much shorter than $t_r/(2\sqrt{LC})$ [3]).

Table 3.2 Global wire parameters

Parameter	Value
Width, W_L (μm)	0.31
Minimum spacing, S (μm)	0.31
Thickness, t (μm)	0.83
Height, h (μm)	0.14
Dielectric constant	2.1
R (Ω/mm)	85.5
Substrate capacitance, C_g (fF/mm)	77.4
Coupling capacitance, C_C (fF/mm)	70.3

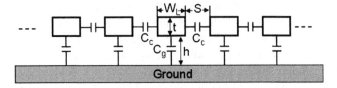

Fig. 3.10 Cross-section of interconnect with wire parameters labeled

3.3.3.1 Comparison with Traditional Two-Inverter Driver

The lookahead system is compared to a traditional two-inverter link driver in Fig. 3.11, with the power savings of the lookahead approach over a traditional two-inverter buffer separated into driver and link transient power consumption, with the peaks at ~1.2 ns corresponding to a rising transition and those at ~1.8 ns corresponding to a falling transition. The power results from Fig. 3.11 were obtained by sizing both systems to meet two voltage swing constraints at the target frequency of 1.5 GHz – each system was sized to pull up to 90% of the link swing voltage (a 0.9 V constraint in the traditional system and a 0.54 V constraint in the lookahead system), and to pull down to 10% of the link swing voltage (a 0.1 V constraint in the traditional system and a 60 mV constraint in the lookahead system). The driver power shown is the amount of power dissipated by four drivers driving a 5 mm, four-bit bus segment with the worst case coupling input (a "1010" pattern in cycle t followed by a "0101" pattern in cycle $t + 1$). The link power shown is the amount of power dissipated by the four-bit bus segment using the parameters from Table 3.2. The peak power in the lookahead approach is 62.7% smaller that of the traditional system, with an average power savings of 48.6%. The above power analysis includes both dynamic and leakage power, with the leakage power further compared in Fig. 3.11b. The use of V_{Low} in the lookahead scheme reduces the output high (V_{OH}) leakage power by nearly 80%, while the reduction in the output low (V_{OL}) leakage – a result of the reduced pull-up network sizing – is ~30%, resulting in an average leakage current reduction of just over 49%.

The power savings of the lookahead approach compared to a traditional approach depend on the clock frequency and activity factor, as shown in Fig. 3.12. In the lookahead approach, the clock frequency controls the amount of time that V_{High} is

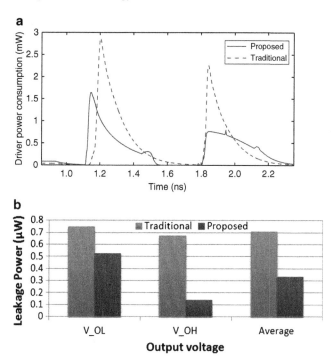

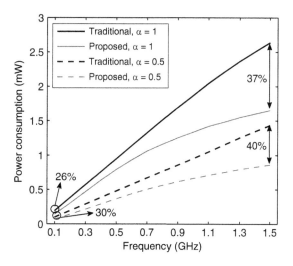

Fig. 3.11 (**a**) Transient power and (**b**) leakage power comparison between the lookahead scheme and the traditional approach

Fig. 3.12 Power comparison of lookahead scheme and traditional scheme with changing frequency and activity factor for a 3-mm link

applied to the link; thus, as the clock frequency increases, the total link swing is reduced, as shown in Fig. 3.13a. This results in power savings of up to 37% for an activity factor $\alpha = 1$. Fig. 3.13b indicates the power consumed at each frequency in the first half of the clock cycle, showing the impact of the changing value of

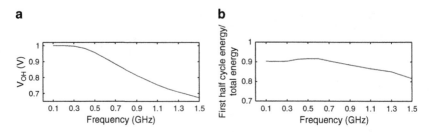

Fig. 3.13 Impact of frequency on (**a**) V_{OH} and (**b**) energy consumption

V_{OH} on the total energy consumption; as frequency increases, the amount of energy consumed in the first half of the clock cycle reduces from 90 to 80%.

Two parameters affect the time that the link remains at logic "1," the clock frequency and the activity factor. Thus, with $\alpha = 1$ and the clock frequency set to 1.5 GHz, the total power savings of the lookahead approach is 37%. If α is reduced to 0.5 for the same clock frequency, the power savings is increased to 40% – V_{High} is only applied for a short period of time, but the link has an additional full clock cycle to reduce all the way to V_{Low}. At very low clock frequencies, the link pulls all the way up to V_{High} and all the way down to V_{Low}, reducing the power benefits to 26 and 30% for $\alpha = 1$ and $\alpha = 0.5$, respectively.

All the previously reported simulations use a link length of 3 mm. Much of the energy savings of the described approach are achieved by only applying V_{High} to the link for a short time period; thus, the system's performance improvements are also tied to the link length. The power dissipation of the described system and a traditional two-inverter buffer system are plotted in Fig. 3.14 for link lengths of 1, 2, and 3 mm. The systems are sized to meet the aforementioned design constraints for the 1.5 GHz, 3-mm link case. In Fig. 3.14a, the power measurements are taken at a frequency of 1.5 GHz, while in Fig. 3.14b the frequency is reduced to 500 MHz.

At each frequency, the power savings of the lookahead approach over the single voltage design decrease as the link length decreases, from a 37% improvement at 3 mm to a 14% improvement at 1 mm. The clock period is fixed, so the reduction in benefit is a result of the reduced load capacitance being charged to a high voltage during the first half of the clock cycle. At the 0.5 GHz frequency, shown in Fig. 3.14b, the power benefits of the lookahead approach are less dependent on link length because the link swing is less affected by the change in load capacitance; the power benefits at 0.5 GHz are between 16 and 20% for link lengths of 3 and 1 mm, respectively.

3.3.3.2   Comparison with Adaptive Voltage Driver

Adaptive voltage systems have been in use for some time in logic circuits and are becoming increasingly popular for improving the energy performance of on-chip interconnect as well [14, 22]. To examine how the benefits of this method would be

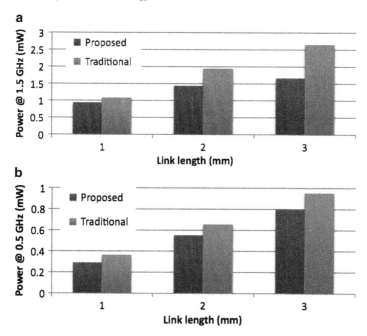

Fig. 3.14 Impact of link length on power dissipation in lookahead scheme and traditional single-voltage scheme (**a**) 1.5 GHz, (**b**) 0.5 GHz

Table 3.3 Power savings of lookahead approach over traditional approach for a range of operating voltages and frequencies

V_{Swing}/V_{High}	System	Target frequency (GHz)				
		1.5	1.25	1.0	0.75	0.5
1.0 V	Trad. (mW)	2.578	2.243	1.872	1.421	0.956
	Prop. (mW)	1.691	1.534	1.364	1.126	0.801
	Savings (%)	34.4	31.6	27.1	20.8	16.2
0.9 V	Trad. (mW)	2.070	1.815	1.525	1.164	0.784
	Prop. (mW)	1.353	1.239	1.112	0.930	0.672
	Savings (%)	34.6	31.7	27.1	20.1	14.3
0.8 V	Trad. (mW)	1.599	1.423	1.211	0.934	0.631
	Prop. (mW)	1.017	0.937	0.855	0.734	0.550
	Savings (%)	36.4	34.2	29.4	21.4	12.8
0.7 V	Trad. (mW)	–	1.051	0.919	0.728	0.498
	Prop. (mW)	–	–	–	0.526	0.423
	Savings (%)	–	–	–	27.7	15.1

affected by an adaptive system, it is compared to a traditional two-inverter system (shown in Fig. 3.2) over a range of operating voltages and frequencies. Both of the designs in Table 3.3 make use of level shifters at the receiver end of the link to reduce short-circuit current in the receiver flip-flop caused by a reduced signal swing. In the traditional two-inverter system, the first inverter is fixed to $V_{DD} = 1$ V, while the second inverter is varied between 1 and 0.7 V, indicated by V_{Swing} in Table 3.3. In the lookahead approach, only V_{High} is adjusted; the internal inverter and the clocked inverter are set to V_{DD}, while V_{Low} is fixed to 0.6 V. Both systems are sized to operate at 90% of their intended link swing at $V_{High} = V_{Swing} = 1$ V and 1.5 GHz, and the voltages and frequencies are then varied using those sizes, as would be the case in an adaptive voltage system.

As shown, the lookahead approach maintains power benefits in excess of 34% at 1.5 GHz until the voltage reduction causes both systems to fail to meet the timing requirement. As the frequency is reduced, the power benefits of the lookahead system are also reduced, as the increased clock period causes the swing voltage to approach V_{High}. The traditional system is also capable of operating at higher frequencies than the lookahead system at 0.7 V; this is a result of the smaller size requirement of the lookahead output driver to meet the 90% swing constraint (90% swing in the lookahead method is just 540 mV, while in the traditional method it is 900 mV). In this comparison, the lookahead method occupies just 1% more area than the traditional method. In addition, as the voltages are scaled down, the average leakage current is further reduced, from a 49% reduction at $V_{High} = V_{Swing} = 1$ V to a 65% average leakage reduction when $V_{High} = V_{Swing} = 0.8$ V.

3.3.3.3 Comparison with Prior Dual-Voltage Switching Method

An alternative method of using two voltages to aid rising link transition performance is described in [29], redrawn below in Fig. 3.15. Rather than the lookahead information used in the proposed scheme, the method shown in Fig. 3.15 uses the value at node Q to apply V_{High} to the output. The system operates as follows. Assume that Q is set low, *Out* is set low, and *In* is set high. The small inverter in the feedback path uses the supply voltage V_{Low} and is unable to fully shut off the PMOS device

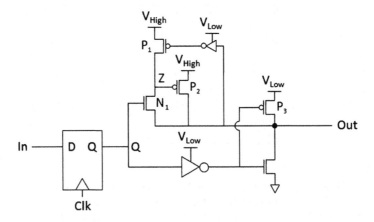

Fig. 3.15 Schematic of prior dual-voltage switching method

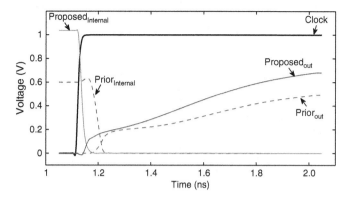

Fig. 3.16 Transient waveform comparison between lookahead approach and prior dual-voltage switching method

P_1 (the gate-source voltage of P_1 is $V_{Low} - V_{High}$). Thus, node Z is charged to V_{High}, which fully shuts off the PMOS device P_2. When the clock is triggered and Q is set high, the NMOS device N_1 is turned on; as node Z is pulled low, P_2 is turned on. The output node will be pulled up through both P_2 and P_3; P_2 shuts off when Out is large enough to toggle the feedback inverter.

The rising delay of the lookahead approach is smaller than the rising delay of the alternative method because the lookahead design bypasses the $Clk{\rightarrow}Q$ delay of the flip-flop, resulting in the faster internal falling signal in Fig. 3.16. In the lookahead approach, V_{High} is also applied for a longer time period, allowing us to reduce V_{High} to just 770 mV and still achieve superior delay performance to the prior system whose V_{High} is set to 1 V, also shown in Fig. 3.16.

Even with V_{High} reduced to 770 mV, the described approach has a latency of less than 66% of the prior approach. The performance of the two systems in terms of power-delay product (PDP) is shown in Fig. 3.17. The lookahead approach is shown to have superior PDP performance across a wide range of frequencies, with an improvement of 42.5% at 1.25 GHz. The improvement decreases to 14.5% at 0.5 GHz; as the clock period is decreased, the lookahead scheme will charge the link closer to V_{High}, reducing the energy benefits.

3.3.4 Limitations

Drawbacks of the lookahead approach include increases in area and complexity. In addition, the use of V_{Low} reduces link reliability by making the link more susceptible to external noise sources. While a number of powerful techniques have been proposed to improve noise tolerance on nanoscale interconnect links [7, 13, 43, 47], common sources of variation including process, voltage, and temperature can still have a dramatic impact on link performance.

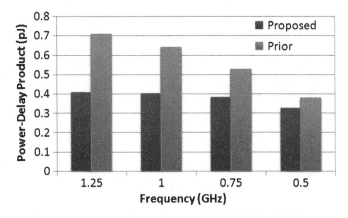

Fig. 3.17 Power comparison between lookahead approach and prior dual-voltage switching method

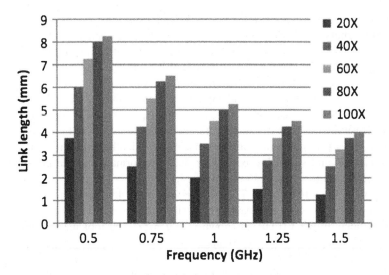

Fig. 3.18 Maximum link length meeting 90% swing constraint at various sizing constraints

An additional drawback of the reduced link swing voltage is that the use of repeater buffers dramatically increases delay compared to a nominal voltage system. The exclusion of repeater insertion techniques is a limitation of the lookahead approach; however, as shown in Fig. 3.18, the lookahead method is capable of driving multi-millimeter link lengths at high frequencies without excessively scaling up the size of the final driver buffer. The multiples of X (20X to 100X) reported in Fig. 3.18 use a 2.5:1 sized inverter for X, with 1 equal to the minimum width of 45 nm. As shown, link lengths of 4 mm (typical NoC link lengths are ≤4 mm [7]) may be driven at 1.5 GHz using driver sizes of 11.25 μm.

The lookahead system achieves significant energy benefits over a fixed V_{High} design without sacrificing delay; furthermore, the lookahead approach requires a much smaller driving buffer to meet the aforementioned constraints. When sized to meet the 10–90% constraints at 1.5 GHz, the area overhead of the lookahead approach is just 1% larger than the traditional approach; however, compared to the prior dual-voltage scheme [29], the lookahead approach has a 33% link area overhead.

References

1. David, J. A., et al.: Interconnect limits on gigascale integration (GSI) in the 21st century. Proc. IEEE **89**, 305–324 (2001)
2. Sridhara, S. R., Ahmed, A., Shanbhag, N. R.: Area and energy-efficient crosstalk avoidance codes for on-chip buses. Proc. IEEE Conf. on Comp. Design (ICCD'04) 12–17 (2004)
3. Ismail, Y. I., Friedman, E. G., Neves, J. L.: Figures of merit to characterize the importance of on-chip inductance. IEEE Trans. Very Large Scale Integration (VLSI) Syst. **7**, 442–449 (1999)
4. Wang, H., Peh, L.-S., Malik, S.: Power-driven design of router microarchitectures in on-chip networks. Proc. 36th IEEE/ACM Int. Symp. on Microarchitecture (MICRO-36), 105–116 (2003)
5. Larsson-Edefors, P., Eckerbert, D., Eriksson, H., Svensson, K. J.: Dual threshold voltage circuits in the presence of resistive interconnects. Proc. IEEE Comp. Soc. Ann. Symp. on VLSI (ISVLSI'03), 225–230 (2003)
6. Jeong, W., Paul, B. C., Roy, K.: Adaptive supply voltage technique for low swing interconnects. Proc. 2004 Asia and South Pacific Design Automation Conf. (ASP-DAC'04), 284–287 (2004)
7. Simone, M., Lajolo, M., Bertozzi, D.: Variation tolerant NoC design by means of self-calibrating links. Proc. Design, Automation and Test in Europe (DATE'08), 1402–1407 (2008)
8. Wei, G. Y., Horowitz, M., Kim, J.: Energy-efficient design of high-speed links. In: Pedram, M., Rabaey, J. (eds.) Power Aware Design Methodologies, Kluwer, Norwell, MA (2002)
9. Akl, C. J., Bayoumi, M. A.: Transition skew coding for global on-chip interconnects. IEEE Trans. Very Large Scale Integration (VLSI) Syst. **16**, 1091–1096 (2008)
10. Weeasekera, R., Pamunuwa, D., Zheng, L.-R., Tenhunen, H.: Minimal-power, delay-balanced SMART repeaters for global interconnects in the nanometer regime. IEEE Trans. Very Large Scale Integration (VLSI) Syst. **16**, 589–593 (2008)
11. Zhang, H., George, V., Rabaey, J. M.: Low-swing on-chip signaling techniques: effectiveness and robustness. IEEE Trans. Very Large Scale Integration (VLSI) Syst. **8**, 264–272 (2000)
12. Benini, L., De Micheli, G., Macii, E., Sciuto, D., Silvano, C.: Asymptotic zero-transition activity encoding for address busses in low-power microprocessor-based systems. Proc. 7th Great Lakes Symp. on VLSI (GLSVLSI'97), 77–82 (1997)
13. Bertozzi, D., Benini, L., De Micheli, G.: Error control schemes for on-chip communication links: the energy reliability tradeoff. IEEE Trans. Computer-Aided Design of Integrated Circuits and Syst. **24**, 818–831 (2005)
14. Worm, F., Ienne, P., Thiran, P., De Micheli, G.: A robust self-calibrating transmission scheme for on-chip networks. IEEE Trans. Very Large Scale Integration (VLSI) Syst. **13**, 126–139 (2005)
15. Ghoneima, M., et al.: Skewed repeater bus: a low power scheme for on-chip buses. IEEE Trans. Circuits and Syst.-I: Fundamental Theory and App. **55**, 1904–1910 (2006)
16. Venkatraman, V., Anders, M., Kaul, H., Burleson, W., Krishnamurthy, R.: A low-swing signaling circuit technique for 65nm on-chip interconnects. Proc. IEEE Int. Soc Conf. (SoCC'06) 289–292 (2006)
17. Stan, M. R., Burleson, W. P.: Bus-invert coding for low-power I/O. IEEE Trans. Very Large Scale Integration (VLSI) Syst. **3**, 49–58 (1995)

18. Shin, Y., Chae, S.-I., Choi, K.: Partial bus-invert coding for power optimization of application-specific systems. IEEE Trans. Very Large Scale Integration (VLSI) Syst. **9**, 377–383 (2001)
19. Yang, J., Gupta, R., Zhang, C.: Frequent value encoding for low power data buses. ACM Trans. Design Automation of Electronic Syst. (TODAES) **9**, 354–384 (2004)
20. Sotiriadis, P., Chandrakasan, A.: Low power bus coding techniques considering inter-wire capacitances. Proc. IEEE Custom Integrated Circuits Conf. (CICC'00) 507–510 (2000)
21. Li, L., Vijaykrishnan, N., Kandemir, M., Irwin, M. J.: A crosstalk aware interconnect with variable cycle transmission. Proc. Design, Automation and Test in Europe (DATE'04) 102–107 (2004)
22. Raghunathan, V., Srivastava, M. B., Gupta, R. K.: A survey of techniques for energy efficient on-chip communication. Proc. Design Automation Conf. (DAC'03) 900–905 (2003)
23. IEEE Standard for Low-Voltage Differential Signals (LVDS) for Scalable Coherent Interface (SCI), 1596.3 SCI-LVDS Standard, IEEE Std 1596.3-1996 (1996)
24. Schinkel, D., Mensink, E., Klumperink, E., Tuijl, E. V., Nauta, B.: Low-power, high-speed transceivers for network-on-chip communication. IEEE Trans. Very Large Scale Integration (VLSI) Syst. **17**, 12–21 (2009)
25. Banerjee, K., Mehrotra, A.: A power-optimal repeater insertion methodology for global interconnects in nanometer designs. IEEE Trans. Very Large Scale Integration (VLSI) Syst. **49**, 2001–2007 (2002)
26. Chen, G., Friedman, E. G.: Low-power repeaters driving RC and RLC interconnects with delay and bandwidth constraints. IEEE Trans. Very Large Scale Integration (VLSI) Syst. **14**, 161–172 (2006)
27. Ghoneima, M., Ismail, Y. I., Khellah, M. M., Tschanz, J. W., De, V.: Reducing the effective coupling capacitance in buses using threshold voltage adjustment techniques. IEEE Trans. Circuits and Syst.-I: Regular Papers **53**, 1928–1933 (2006)
28. Lin, Y., He, L.: Dual-Vdd interconnect with chip-level time slack allocation for FPGA power reduction. IEEE Trans. Computer-Aided Design of Integrated Circuits and Syst. **25**, 2023–2034 (2006)
29. Kaul, H., Sylvester, D.: A novel buffer circuit for energy efficient signaling in dual-VDD systems. Proc. 15th ACM Great Lakes Symp. on VLSI (GLSVLSI'05) 462–467 (2005)
30. Deogun, H. S., Senger, R., Sylvester, D., Brown, R., Nowka, K.: A dual-VDD boosted pulsed bus technique for low power and low leakage operation. Proc. IEEE Symp. Low Power Electronics and Design (ISLPED'06) 73–78 (2006)
31. Wang, P., Pei, G., Kan, E. C.-C.: Pulsed wave interconnect. IEEE Trans. Very Large Scale Integration (VLSI) Syst. **21**, 453–463 (2004)
32. Jose, A. P., Patounakis, G., Shepard, K. L.: Near speed-of-light on-chip interconnects using pulsed current-mode signalling. Proc. Symp. VLSI Circuits 108–111 (2005)
33. Khellah, M., Tschanz, J., Ye, Y., Narendra, S., De, V.: Static pulsed bus for on-chip interconnects. Proc. Symp. VLSI Circuits 78–79 (2002)
34. Katoch, A., Veendrick, H., Seevinck, E.: High speed current-mode signaling circuits for on-chip interconnects. Proc. IEEE Int. Symp. Circuits and Syst. (ISCAS'05) 4138–4141 (2005)
35. Bashirullah, R., Wentai, L., Cavin, R., III: Current-mode signaling in deep submicrometer global interconnects. IEEE Trans. Very Large Scale Integration (VLSI) Syst. **11**, 406–417 (2003)
36. Kumar, S., et al.: A network on chip architecture and design methodology. Proc. IEEE Comp. Society Ann. Symp. on VLSI (ISVLSI'02) 105–112 (2002)
37. Amde, M., Felicijan, T., Efthymiou, A., Edwards, D., Lavagno, L.: Asynchronous on-chip networks. IEE Proc. Comput. Digit. Tech. **152**, 273–283 (2005)
38. Bjerregaard, T., Sparso, J.: A router architecture for connection-oriented service guarantees in the MANGO clockless network-on-chip. Proc. Conf. Design, Automation and Test in Europe (DATE'05) 1226–1231 (2005)
39. Lines, A.: Asynchronous interconnect for synchronous SoC design. IEEE Micro **24**, 32–41 (2004)
40. Rosenfeld, J., Friedman, E. G.: Quasi-resonant interconnects: a low power, low latency design methodology. IEEE Trans. Very Large Scale Integration (VLSI) Syst. **17**, 181–193 (2009)

41. Shang, L., Peh, L.-S., Jha, N. K.: Dynamic voltage scaling with links for power optimization of interconnection networks. Proc. Int. Symp. High Perf. comp. Arch. (HPCA'03) 91–102 (2003)

42. Kaul, H., Sylvester, D., Blaauw, D., Mudge, T., Austin, T.: DVS for on-chip bus designs based on timing error correction. Proc. Design, Automation and Test in Europe (DATE'05) 80–85 (2005)

43. Fu, B., Ampadu, P.: On Hamming product codes with type-II hybrid ARQ for on-chip interconnects. IEEE Trans. Circuits and Syst.-I: Regular Papers **56**, 2042–2054 (2009)

44. Fu, B., Wolpert, D., Ampadu, P.: Lookahead-based adaptive voltage scheme for energy-efficient on-chip interconnect links. Proc. 3rd ACM/IEEE Int. Symp. on Networks-on-Chip (NoCS'09) 54–64 (2009)

45. Arizona State University, Predictive Technology Model [Online]. Available: http://www.eas.asu.edu/~ptm/

46. Wong, S., Lee, G., Ma, D.: Modeling of interconnect capacitance, delay, and crosstalk in VLSI. IEEE Trans. Semiconductor Manufacturing **13**, 108–111 (2000)

47. Xu, S., Benito, I., Burleson, W.: Thermal impacts on NoC interconnects. Proc. IEEE Int. Symp. on Networks-on-Chip (NoCS'07) 220–220 (2007) Full version Available: http://python.ecs.umass.edu/ icdg/publications/pdffiles/xu_noc07.pdf

Chapter 4
Asynchronous Communications for NoCs

Stanislavs Golubcovs and Alex Yakovlev

Abstract Technology scaling beyond 90 nm drastically complicates the chip design process. Greater demand for higher performance and more functionality placed on a single chip, while maintaining power consumption at a reasonable level drives research towards new architectural and communication paradigms that support topological scaling. Network-on-Chip is seen as one such paradigm, but its inherently massive parallelism and distribution of switching activity naturally lead to a much wider spectrum of techniques used for system timing. The use of global clocking becomes very difficult for improving power and performance while at the same time keeping acceptable levels of robustness to faults, both fabrication and run time, as well as to the increasing parametric variability of components.

Systems based on NoCs are thus becoming more diverse in terms of timing, and if not fully asynchronous, then mixed, e.g. globally asynchronous and locally synchronous. The notion of timing and synchronization is pervasive in system communication architectures and affects all layers of hierarchy, but its biggest effect is probably at the link layer, where the notion of data validity in communication channels between processing nodes and network routers is paramount. This chapter provides an overview of the various asynchronous techniques that are used in such links, including signalling schemes, data encoding and synchronization solutions. Those are discussed with a view of comparison in terms of area, power and performance. The fundamental issues of the formation of data tokens based on the principles of data validity, acknowledgement, delay-insensitivity, timing assumptions and soft-error tolerance are considered. The chapter also covers some of the aspects related to combining asynchronous communication links to form parts of the entire network architecture, which involves asynchronous logic for arbitration and routing hardware. To this end, we also present basic techniques for building small-scale controllers using the formal models of Petri nets and signal transition graphs.

A. Yakovlev (✉)
Asynchronous Systems Laboratory, School of EECE, Newcastle University,
Newcastle upon Tyne, United Kingdom
e-mail: Alex.Yakovlev@ncl.ac.uk

C. Silvano et al. (eds.), *Low Power Networks-on-Chip*,
DOI 10.1007/978-1-4419-6911-8_4, © Springer Science+Business Media, LLC 2011

4.1 Introduction

At a broad overview, we can say that the communication between various devices
can be organised in two opposing approaches (Fig. 4.1). One is to create a dedicated
point-to-point link between each pair of components. Another is to provide some
common communication medium such as bus. Both methods have a limiting factor
as the number of interconnected devices grows. In the first case, the number of
links increases quadratically and soon enough the designer faces a massive overhead of links being created. In the second case, as more and more devices may
need to transmit data, the availability of the communication medium shrinks, which
limits the potential throughput. The Network-on-Chip architecture is a hybrid of
these concepts. It organizes multiple point-to-point links in order to increase the
communication parallelism while reusing the same links to communicate between
different cores.

Design of NoCs is very complex as there is a great variety of design options and
many aspects and criteria of quality. This complexity becomes even greater when the
global clock is removed. Among the rest of the network components (Fig. 4.2), the
communication links pose a variety of possible signalling schemes while offering a

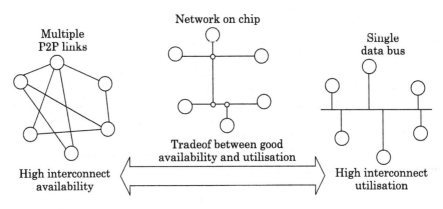

Fig. 4.1 P2P links vs. bus

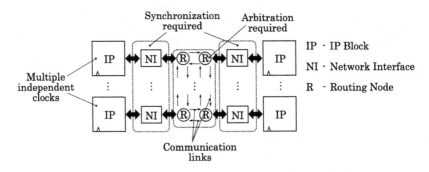

Fig. 4.2 Generic NoC structure

wide range of encoding methods, which in turn multiplies the number of possible implementations. This requires a good classification scheme that addresses possible forms of signals for transmitting data (Fig. 4.4).

A token-based approach allows us to unify some of those categories of communication mechanisms. It helps us to formulate, first of all, a reliable form of communication action. On top of this, we should also consider the criteria according to which we can compare different solutions, such as performance (throughput), power consumption and area.

The basic difference between synchronous and asynchronous designs lies in the use of the global clock signal for updating the state of the system. Asynchronous design does not have the global clock, and hence it is sometimes called clockless or self-timed. Instead, its function is based on many, often mutually concurrent, control signals causing partial system state updates. Asynchronous circuits are typically classified as [53]:

- Delay-insensitive circuits, where the circuit is expected to work correctly regardless of wire and gate delays
- Speed-independent circuits, where the circuit is expected to work correctly regardless of gate delays
- Circuits with timing assumptions, where the circuit is expected to work correctly if certain race conditions are met

Technically, synchronous logic can be regarded as asynchronous with the only control signal being the clock, and the timing assumption stating that each possible signal propagation path between two state registers will take no longer than a certain predefined period of time. The longest propagation time is identified by finding the longest path between the registers storing the state of the system. This path is called the *critical path* and is essentially formed by the delays of the gates, or cells, along this path and the delays of the wires interconnecting these gates.

4.1.1 Variability

As gates are getting smaller, their associated delay and local interconnects are getting smaller. On the contrary, the delay of long (usually, system-level) wires is magnified exponentially and has a much greater impact on the critical paths and the overall performance of the system (Fig. 4.3). This requires the designer to introduce additional repeaters that would reduce the overall latency of such links. More than that, shrinking of the layout feature sizes causes an increase in interconnect delay variability for both traditional die-to-die and emerging intra-die variations [66]. The delay variability also causes clock skew variations. For the $0.25\,\mu\text{m}$ process, the reported variability was already 25% [27]. And finally, the crosstalk capacitance can cause up to $7\times$ delay variation [17]. All these factors require much more sophisticated timing analysis. Safety margins associated with the variability have to be increased at the cost of system performance.

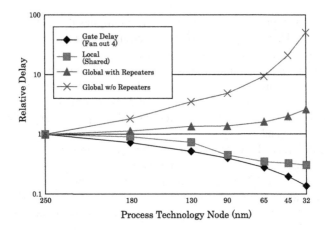

Fig. 4.3 Delay for local and global wiring vs. feature size [18]

From the perspective of the clockless design, similar issues occur. However, the circuit is generally more immune to variations in gate and wire delays. In the common case, there are no timing constraints on the allowed delay of the longest path. Instead, there are request and acknowledge handshake interfaces connecting separate, more or less independent, circuit parts. These produce separate communication transactions. Typically, each transaction consists of three sequential phases:

- The request phase – initiates the new transaction.
- The acknowledge phase – signals that the request was received and the associated operation was accomplished.
- The reset phase – with some protocols (such as 4-phase or return-to-zero). The reset phase is necessary to initiate the signals and prepare for the next transition.

The estimation of the performance of such a circuit is based on how frequently new transactions (new requests) may be sent from the sender to the receiver. For obvious reasons, the acknowledge and the reset phase should be included in the estimation. The loops forming the requests can be considered as chains of handshakes. The overall performance of such a chain is the average latency of the statistically slowest loop in the system.

4.1.2 Power Consumption

Another important aspect of asynchronous design is its prudence towards circuit power consumption. In synchronous systems, the clock signal needs to be propagated across the whole chip. The associated effort to drive all these wires within tolerable skew requires significant amount of power. The asynchronous design is based on more or less localised handshake interconnects. As a result, it often requires much less power to function.

Another advantageous aspect is event-based power consumption. In a synchronous system, the clock signal drives each bistable element of the circuit irrespective of real data changes. On the contrary, the asynchronous design is driven by the "on demand" philosophy. It dissipates power only when it is activated by the incoming request signal. Significant power savings in clocked design can be achieved by gating the clock for inactive circuit regions; however, implementing it requires additional area and design effort.

Static power consumption is related to the technology used and the area of the circuit. For CMOS technology of 130 nm and above, static power consumption is a minor contributor. As it scales down to 90 nm and beyond, static (leakage) power becomes a greater concern. Again, asynchronous techniques allow applying power gating in a more flexible event-based way than in clocked circuits [26].

4.1.3   Chapter Structure

The main goal of this chapter is to provide an overview on a variety of issues associated with designing NoC links (Fig. 4.4). The chapter starts with explaining the mechanisms of asynchronous communication, followed by the main link-level protocol issues, such as data flow control schemes, data-encoding techniques and error resilience. After that the chapter moves to the various implementation issues that need to be considered when multiple links are formed into networks. Here, the ways of improving throughput, power and area, by means of pipelining, serialisation and acknowledgement hiding are considered. Finally, the chapter mentions some design methods that can be used when creating asynchronous links and shows one simple example of designing a link controller. For a more practically oriented reader, we hope that this material can also be used as a guide into the design of on-chip asynchronous communication links, to help the designer select the appropriate protocol, style of implementation and circuit design method.

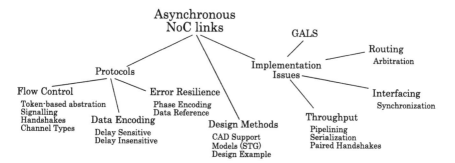

Fig. 4.4 Chapter overview

4.2   History of Asynchronous Communications Before the NoC Era

Historically, needs for asynchronous interfaces have always been a strong motivating factor for research and development in the area of self-timed systems. Simple system designs initially required to have a single controller device communicating with a detachable testing device or a processor communication with peripheral devices such as monitor and printer. People later realised the necessity to unify the way the devices are interconnected. As a result, many link level standards (such as shared buses) had emerged and kept evolving over time improving a variety of the link characteristics such as speed, power consumption and area. As it was natural for the modules to be large in size and different in response time, clocking them with a single signal would be very difficult if not impossible, and in most occasions it was natural to use asynchronous communication techniques. They also used asynchronous arbitration in order to allow different masters to use the same bus to access slaves. Among the industrial examples of such buses that appeared from 1960s to 1980s were IEEE-488/HP-IB/GPIB, UNIBUS (from DEC), LSI-11/Q-Bus, VME-Bus (from Motorola), Futurebus/IEEE-896, which all had asynchronous handshake protocols and some products had asynchronous logic for the implementation of their adaptors. At the dawn of the VLSI technology (1979), a self-timed interface TRIMOSBUS was introduced [57], which had an elegant way of using three wires (shifting the role of request and acknowledgement after each data transfer and using wire-OR logic on the wires) for one-to-many (broadcast) handshake communications. A range of interesting handshakes, including those that allowed sharing wires for sending request and acknowledgement, as well as acknowledging data on the individual bit lines were introduced in [63].

At some point, people realised that bus did not scale well and could be a point of vulnerability in safety-critical applications, such as on-board multi-computer systems. To address this issue, a self-timed communication channel with a ring architecture was developed in 1986 by the group of V. Varshavsky and presented in [64, 68]. This was probably the first kind of a globally asynchronous locally synchronous system, where synchronous processors, for the purposes of design re-use and code legacy, could communicate through the ring-based fault-tolerant communication network. The processors were interfaced to the ring channel via special synchronous–asynchronous interfaces, self-timed FIFOs and self-timed routers. The ring channel used an original token-based protocol with dynamic priority arbitration, multi-cast addressing and links that used return-to-zero handshakes with 3-of-6 Sperner encoding. The channel adaptors, built in a fully speed-independent way, and wire redundancy in the links, allowed automatic detection, location and recovery (by reconfiguration) up to two stuck-at faults in link wires and in the gates inside the router logic.

In many ways, current developments in the area of asynchronous on-chip communications follow those prior developments adding new encodings, circuit variations, changed design criteria and, above all, using CAD tools of the day.

4.3 Token-Based View of Communication

At a high level, the data communication of two or more elements can be modelled as
a token-based structure. Consider the link transporting data between the transmitter
and the receiver (Fig. 4.5a). The following two basic questions have to be resolved:

- (Q1) When can the receiver look at the data?
- (Q2) When can the transmitter send new data?

The first question is related to the *data validity issue*. When the data are available a
well-defined token is formed (Fig. 4.5b). The second question regards the *acknowledgement issue*. Essentially, it is a feedback from the receiver that helps separating
different tokens and creating the dynamics of the data transmission.

These are the fundamental issues of flow control at the physical and link levels.
The answers to these questions are determined by many design aspects: technology,
system architecture (application, pipelining), latency, throughput, power, design
process, etc.

Using Petri net models [35], we can describe the data communication process
in a sufficiently abstract and formal way (Fig. 4.6). Places "data valid" and "ack"
would interpret the mutually exclusive states. First, it would tell the receiver when
the new data are available and then inform the transmitter that the data were received
and the new data may be transferred at any time.

Such an abstraction can provide a simple data flow view of a discrete process.
Without too many details, these models can describe arbitrary network structures
and dynamically visualise the computation or the communication process through
the token game simulation. As far as the token game, such a model can be constructed and simulated using existing visual tools [44], but what about its actual
implementation? What does it mean to "produce" or "consume" a token in our communication context?

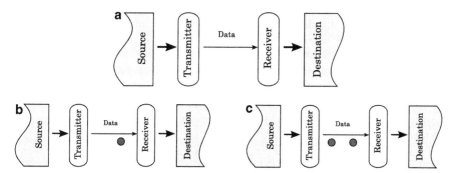

Fig. 4.5 Token-based view of communication. (**a**) Data link (**b**) Token signifies the presence of
valid data (**c**) Next token has to be separated

Fig. 4.6 Link modelled
using Petri nets

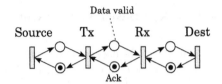

When we speak about implementing such a model on a silicon wafer, we can imagine that the links that are used for transporting data have to be based on metal wires connecting the modules. The tokens are somehow encoded by the current flowing through these wires or by the voltage levels on them.

There are a number of limitations in the physical world. The first question is related to the problem of detecting when the actual data are available. It is also related to the so-called completion detection problem.

The wire by itself is not a medium that can store multiple tokens at a time. In digital design, it is also limited to only binary values which it can transport. That requires using a group of wires representing the same link and essentially extends the variety of implementations with different functional characteristics.

For technical reasons, the receiver may require some time before the transmitter sends the next portion of information. It may happen because either the receiver is occupied processing the previous token or there is a congestion and the destination module is temporarily unable to accept new tokens from the receiver. Hence, before sending a token, the transmitter needs to know that the receiver will be able to process it. Hence, something has to be done to signify the availability of the receiver. The actual implementation of any communication link would contain an answer for each of these questions in both synchronous and asynchronous designs.

Often, the readiness of the receiver to receive the next datum is based on the timing assumptions. One example is the Internet routers that transmit the data over a single wire. Then, special transition sequences are used to encode both the timing of the token and its value. The ultimate assumption is that the rate at which the tokens are sent is slow enough for the receiver to absorb them.

In traditional globally clocked systems, the time of the token arrival is identified by the common clock signal [21]. The actual data value is binary encoded over a number of wires. And finally, the acknowledgement is implied by the assumption that the signal has enough time to propagate and get processed on the receiver side. Note that the common clock is making a common token event for each receiver in the circuit. This means that each link will transmit exactly one token per cycle. In most cases that would require additional validity bit to signal whether the value of the token can be used.

Finally, for partially clocked or pure asynchronous systems, data communication relies on the notion of local handshakes. Some of them are based on timing assumptions (e.g. bundled data) while others are delay-insensitive (dual-rail, 1-of-4). Each of these approaches assumes two-way communication, where the propagation of the request signal from one party is followed by the acknowledgement of the other party.

In the next sections we will consider basic concepts and the related techniques for the formation and separation of tokens. They cover signalling protocols and special encoding techniques to transmit data and validity information.

4.4   Basics of Asynchronous Signalling

4.4.1   Signalling Techniques

There are several signalling techniques that can be used to implement the events of a token-based model in order to organise asynchronous communication (Fig. 4.7). The most common are *level signalling* and *transition signalling*.

The level signalling encodes an event by assigning a particular signal voltage level to the wire. The concept is closely related to the way signals are represented on wires in logic circuits and in practice is simple to implement using logic gates. The drawback is the requirement at some point to reset the signal to its initial state. In other words, one event in the modelled behaviour would cause two signal transition events in the actual implementation. This makes such an approach much less attractive when dealing with long link communications, where each signal transition is considered to be costly.

The alternative, transition signalling, is based on switching the voltage level on the wire to the opposite when an event occurs. Such a correspondence is more attractive from the modelling point of view. However, the transitions are different (either raising or falling edge) and current value of the wire cannot be used to detect the event unless additional logic gates are used. To detect that a signal transition has occurred, the receiver needs to compare the signal value to some reference signal, which has remained unchanged, or with the previous value, which often needs to be stored.

The less common technique is based on *signal pulses*. Its advantage is a combination of simplicity from the level-based signalling with the close correspondence between a modelled event and an event on wires. However, this approach requires careful maintenance of the pulse width which strongly depends on physical parameters of the associated circuitry. The technique is particularly interesting and is more closely considered in one of the following sections.

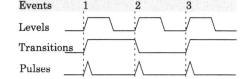

Fig. 4.7 Signalling
techniques

4.4.2 Handshake Protocols

Handshakes are the backbone of asynchronous communications. They provide the flexibility of composing independent modules into communicating systems. A handshake consists of two basic events: the request event initiating the transaction and the acknowledgement event finalising it.

Depending on the signalling scheme used, a handshake can either be 4-phase (Fig. 4.8a) or 2-phase (Fig. 4.8b).

The 4-phase handshake is related to the level-based signalling. It is also commonly known as a return-to-zero (RTZ) method. It forms the sequence of the following events:

Phase 1   *Req* ↑ – A request is issued, new communication cycle has started.
Phase 2   *Ack* ↑ – The request is acknowledged. The requester may proceed with the transaction.
Phase 3   *Req* ↓ – The requester is resetting to the initial state and waiting for the responder to reset as well.
Phase 4   *Ack* ↓ – The initial state of the handshake is restored and can process to Phase 1.

The 2-phase handshake can be related to the transition or the pulse-based signalling. It has only two phases per handshake:

Phase 1   Either *Req* ↑ or *Req* ↓ – A request is issued and the new communication cycle has been started.
Phase 2   Either *Ack* ↑ or *Ack* ↓ – The request is acknowledged, can proceed to Phase 1.

Note that when both *Req* and *Ack* signals are used, the event detection on both sides can use these signals as a reference. If *Req* ≠ *Ack*, the responder knows that there was an event on the request line. If *Req* = *Ack*, then the requester knows that the acknowledgement has been sent.

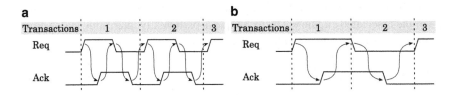

Fig. 4.8 Basic handshake protocols. (**a**) 4-phase handshake (**b**) 2-phase handshake

4.4.3   Channel Types

The type of the channel may differ depending on who the initiator of the transmission is. When the transmitter is making the initial request, it forms the so-called push channel as the data being sent are associated with the request event (Fig. 4.9a).

In the opposite situation, the receiver may also activate the transmission. Such communication would be referred as the *pull channel* because the receiver is the one requesting for data (Fig. 4.9b). The sender then acts as a responder with the acknowledgement event delivering the data.

For the full picture, we may imagine that there is no data transmission in the channel, only the handshake to communicate control information. Then it would be called the *nonput channel*. It does not transport the data but can still be useful to synchronize the modules.

Finally, both the transmitter and receiver may also attach some information to the events. Such a communication link is called the *biput channel* [53].

Since there is such a variety of the data transmission directions, it is sometimes also convenient to call the participants the active side (or the master) initiating the request and the passive side (or the slave) responding to the initial request with the acknowledgement.

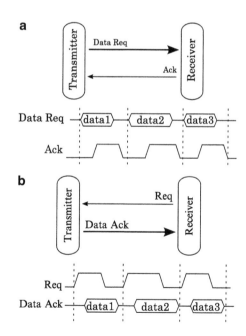

Fig. 4.9 Channel types.
(a) Push channel handshake
(b) Pull channel handshake

4.5 Delay-Insensitive Data Communication

Delay-insensitive codes [65] allow the signal propagation time for each separate wire and for each separate communication to be different. It is only required that within a finite amount of time the signal arrives to the destination.

Classification factors describing these codes are the following:

- Efficiency – the rate at which A different codewords can be encoded over the N wires:
$$R = \frac{\lfloor \log_2 A \rfloor}{N}.$$
- Completion detection – complexity of data validity detection.
- Encoding/decoding – complexity of converting the between codes that are used in the channel and inside the modules.

Delay-insensitive techniques encode both the event (some indication that the input represents fresh and valid data) and the associated data into the same wires. Special *completion detection* circuitry is used at the receiver to identify when the data are available. The complexity of such a circuitry can be an important factor when choosing the communication method.

4.5.1 Dual-rail

Dual-rail logic uses two separate wires ($x.false$ and $x.true$ or $x.0$ and $x.1$) to transmit zeros and ones. In a traditional 4-phase handshake, only one wire can be active at a time. Hence, there are three allowed combinations (Table 4.1).

Activating one of the signals will effectively transmit either 0 or 1. Activating both signals is not allowed, and when a sequence of data symbols is transmitted using a 4-phase protocol, they must be separated by spacers.

Several dual-rail pairs can be grouped to form a multi-bit channel. The n-bit value requires to use $N = 2n$ wires and can accommodate 2^n different codewords.

Figure 4.10a demonstrates an example of push channel communication for two bits using dual-rail codes. The completion detection is organised using n 2-input OR gates. They detect a valid data on each of the bit channels. Then, all validity signals are combined into one completion signal using a C-element [48] (Fig. 4.10b). This component is a latch that outputs the value of the inputs when these inputs match and preserves its value otherwise. In general, for the n-bit channel a tree of two-input C-elements can be used.

Table 4.1 Dual rail 4-phase codes	$x.0$	$x.1$	meaning
	0	0	spacer
	1	0	send 0
	0	1	send 1
	1	1	not allowed

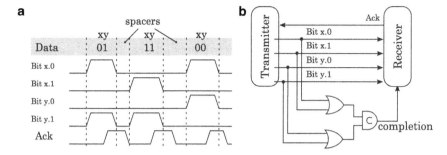

Fig. 4.10 Dual-rail encoding. (**a**) Communication wave forms for the 2-bit transfers (**b**) Completion detection

Note that the concept of spacers does not always imply the use of 0 values on the dual-rail channel lines. The negative logic optimization might provide a better circuit by inverting one or both ($x.0$ and $x.1$) rails. Also, the dual-rail channel may have the alternating spacer values, thus providing a more balanced power consumption in security aware applications [49].

4.5.2   1-of-N and M-of-N

The 1-of-N (also known as one-hot) encoding is used to encode N different symbols over N wires where one wire is dedicated to each of the symbols. The completion detection for such an encoding is trivial. It is an *OR* function of all signals. Such an encoding is very power efficient as transmitting any codeword requires only a single transition, and then the spacer is encoded by a transition on the same wire.

1-of-N is less practical than the dual-rail in general case, as for the increasing N the implementation of the channel quickly becomes infeasible. Yet, the 1-of-4 data encoding is an exception because it also uses four wires to encode 2 bits while it needs fewer transitions. As a result, N-bit data channels can be constructed of $N/2$ four-pin links which would be better than the corresponding N dual-rail links.

The considered 1-of-N encoding is a special case of M-of-N encoding. The logical generalisation would be to use some fixed number of active wires to encode the data. Activating M wires out of N enables us to encode $\binom{N}{M} = \frac{N!}{M!\cdot(N-M)!}$ different values. As an example, 2-of-4 code is able to provide six different codewords over four wires. For N wires, the maximum number of combinations can be obtained, when $N = 2M$ (these are also called the Sperner Codes [62, 65]).

Using Stirling's approximation:

$$\lim_{n \to \infty} \frac{n!}{\sqrt{2n\pi}n^n e^{-n}} = 1, \tag{4.1}$$

Fig. 4.11 Completion
detection. **(a)** 2-of-4
completion detection
(b) Incomplete 2-of-7
completion detection

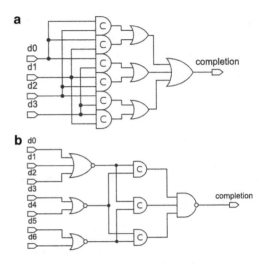

it can be shown that in the best case the total number of possible codewords is:

$$\binom{N}{M} = \binom{2M}{M} = \frac{(2M)!}{(M!)^2} \approx \frac{\sqrt{4M\pi}(2M)^{2M}e^{-2M}}{2M\pi M^{2M}e^{-2M}} = \frac{2^{2M}}{\sqrt{\pi M}}. \qquad (4.2)$$

The completion detection can be constructed of various combinations of C-elements and *OR* gates [4]. Figure 4.11a shows one such completion detector for the 2-of-4 codes. Unfortunately, with growing M and N the complexity of the supporting circuitry grows exponentially. One way to get the benefit of the N-of-M coding while avoiding the excessive complexity of the completion detection is by using the *incomplete N-of-M codes* [4]. For instance, the 2-of-7 coding provides 21 codewords over seven wires. It can encode up to 4 bits of information while leaving five codewords unused. Instead, 2-of-7 can be composed of 1-of-3 and 1-of-4. A composition of these encodings provides 12 different code words. Additionally, 2-of-4 encoding can be added for the case when the first 3 bits are 0. This adds up six more useful combinations (Fig. 4.12a). The combinations when two of the first 3 bits are active were not included. This leaves only 18 useful codewords which still can store 4 bits of information plus two control symbols. The implementation has shown about 25% improvement over the complete 2-of-7 code [4].

We can also present a variant of 3-of-6 code [62] and the relatively simple decoding to binary logic (Fig. 4.12b). The implementation is straightforward, using majority gates:

$$y_1 = p_2; \ y_2 = p_2,$$

$$y_3 = p_1p_3 + p_2p_5 + p_3p_5,$$

$$\hat{y}_3 = p_1p_6 + p_2p_4 + p_4p_6,$$

a

1-of-3	1-of-4	Binary value
1 0 0	0 0 0 1	1 1 0 0
1 0 0	0 0 1 0	1 1 0 1
1 0 0	0 1 0 0	1 1 1 0
1 0 0	1 0 0 0	1 1 1 1
0 1 0	0 0 0 1	1 0 0 0
0 1 0	0 0 1 0	1 0 0 1
0 1 0	0 1 0 0	1 0 1 0
0 1 0	1 0 0 0	1 0 1 1
0 0 1	0 0 0 1	0 1 0 0
0 0 1	0 0 1 0	0 1 0 1
0 0 1	0 1 0 0	0 1 1 0
0 0 1	1 0 0 0	0 1 1 1
0 0 0	1 0 0 1	0 0 1 0
0 0 0	1 0 1 0	0 0 1 1
0 0 0	1 1 0 0	←
0 0 0	0 1 0 1	0 0 0 0 ⎱ unused
0 0 0	0 1 1 0	0 0 0 1 ⎰ symbols
0 0 0	0 0 1 1	←

2-of-4

b

n	p_1 p_2 p_3 p_4 p_5 p_6	y_1 y_2 y_3 $\hat{y}_3$ y_4 $\hat{y}_4$
s	0 0 0 0 0 0	0 0 0 0 0 0
00	0 0 0 1 1 1	0 0 0 1 0 1
01	0 0 1 1 0 1	0 0 0 1 1 0
02	0 0 1 0 1 1	0 0 1 0 0 1
03	0 0 1 1 1 0	0 0 1 0 1 0
04	0 1 0 1 0 1	0 1 0 1 0 1
05	0 1 1 1 0 0	0 1 0 1 1 0
06	0 1 0 0 1 1	0 1 1 0 0 1
07	0 1 1 0 1 0	0 1 1 0 1 0
08	1 0 0 0 1 1	1 0 0 1 0 1
09	1 0 0 1 0 1	1 0 0 1 1 0
10	1 0 1 0 1 0	1 0 1 0 0 1
11	1 0 1 1 0 0	1 0 1 0 1 0
12	1 1 0 0 0 1	1 1 0 1 0 1
13	1 1 0 1 0 0	1 1 0 1 1 0
14	1 1 0 0 1 0	1 1 1 0 0 1
15	1 1 1 0 0 0	1 1 1 0 1 0
16	1 0 1 0 0 1	1 0 1 1 0 0
17	1 0 0 1 1 0	1 0 0 0 1 1
18	0 1 1 0 0 1	0 1 0 0 1 1
19	0 1 0 1 1 0	0 1 1 1 0 0

Fig. 4.12 M-of-N examples. (**a**) Incomplete 2-of-7 code (**b**) 3-of-6 code [62]

$$y_4 = p_1 p_4 + p_2 p_3 + p_3 p_4,$$
$$\hat{y}_4 = p_1 p_5 + p_2 p_6 + p_5 p_6.$$

Here the first 2 bits are single buffers. The other 2 bits are represented in dual-rail. This code and the above logic were used in designing a self-timed fault-tolerant ring channel described in [64, 68].

4.5.3 Single Transition Codes

To some extent, the previously considered 1-of-4 signal encoding can be viewed as a generalisation for the single-rail and dual-rail 4-phase protocols. Similarly, a more generic form can be constructed for the 2-phase protocols. We can define such a protocol as a number of wires $N = 2^n$ transmitting n bits of information by making only a single transition per transfer.

For $n = 0$, it is a single wire transmitting the request signal alone which is the traditional 2-phase request signal.

For $n = 1$, there are two data wires transmitting one bit of data. The first wire signal is changed when the transmitted data bit is switched to the opposite. The second wire is changed when the same bit is being transmitted in the next transaction. This is the so-called level-encoded dual-rail (LEDR) protocol transporting one bit of information per transition [13].

For $n = 2$, there are four data wires transmitting 2 bits of information per transaction. It is also called 1-of-4 level-encoded transition-signal (LETS), described in more detail in [30]. Each of the four codewords ($S_1...S_4$) can be placed on a 4-dimensional hypercube nodes. Each edge in such a 4-cube would represent a single transition on one of the connection wires. Each 2-bit codeword is associated with four neighbouring nodes in such a way that every other codeword can be reached by passing one edge. One such assignment is shown in Fig. 4.13. The data wires a, b, or c are switched when the next codeword differs from the last sent. A dedicated "parity" wire d signifies the repeated transfer of the same codeword.

The common property of all single transition codes is the fact that for each transition the parity of the number of activated wires is always changing. It is also depicted on the diagram by the black and white nodes. This fact allows the completion detection to be done by *XOR*-ing all of the data wires.

The protocol has the disadvantage of exponential growth of connecting wires required to support it. Hence, it may not be practical for $N > 4$.

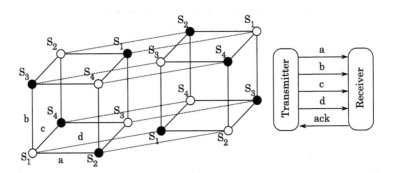

Fig. 4.13 Hypercube encoding 2-bit data transitions

4.6 Delay-Sensitive Communication

Delay-sensitive communication allows the introduction of assumptions that certain concurrent events would always happen in a particular expected order. By assuming so, it is often possible to make the circuit simpler and potentially improve its performance. However, additional care must be taken to make sure that such an assumption is true. As a result, the design has to be approached at a lower level requiring additional timing analysis and transistor level or analog implementation.

4.6.1 Bundled-Data Encoding

The *bundled-data* signal encoding is one of the most popular data transmission techniques. It is considered fast, simple and practical. The term bundled-data refers to a setup when the data symbol is encoded using a conventional binary number notation, and an additional event signal is used to signify when the data are valid. It means that the n-bit encoding requires only $N = n + 1$ wires. The timing assumption is applied here, stating that by the time the request signal approaches the receiver, all of the data signals will be stabilised and usable. As a result, the completion detection circuitry also becomes trivial – only needs to check what the request value is.

Figure 4.14 demonstrates the data wires bundled with *Req* and forming the 4-phase push channel. The data wires need to be settled by the time the request event reaches the receiver. There are possibilities to form different handshakes and channel types. For instance, the bundled-data can also be 2-phase (e.g. to reduce timing overheads in the channel) as shown in [54].

Unfortunately, the bundled data encoding may also have its flaws. While it is possible to roughly estimate the delay of a given link and also control it by varying the size of the driving transistors [56], the increasing variability for smaller technology would require introducing bigger safety margins and subsequently cause degradation of performance.

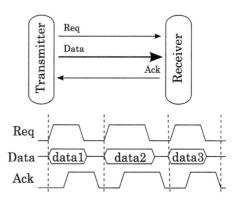

Fig. 4.14 Bundled data transmission

4.6.2 Single-Track Signalling

The single-track handshake protocol uses the same wire to issue both the request and the acknowledgement [61]. Just like other protocols, it can work with both push and pull channels. Essentially, the initiator of the communication sets up the request by raising the level of the wire signal. After some time, the second party removes (or "consumes") the signal returning the wire back to the initial state. As both sides are able to sense the current level on the wire, the communication is established in both directions. This way the protocol combines the advantages of both the 2-phase and the 4-phase protocols. Just like in the 2-phase protocol, it has only two transitions per transaction. And at the same time, each of its transactions starts from a predefined state leading to a simpler completion detector.

The communication can be supported by the data bundled with this single wire. One disadvantage is the necessity of using non-standard cells which may otherwise result in loss of robustness in the communication channel.

4.6.3 Pulse-Based Signalling

An alternative way to encode events on wires is based on producing voltage pulses (Fig. 4.15b). Similarly to the 2-phase coding, there are only two pulses per handshake. Hence, such events can be easily associated with the events in a respective Petri net model. As opposed to the 2-phase signalling, it does not have two different signal levels representing the same event. As a result, the system does not need to explicitly deal with as many states and can have a simpler circuit to implement it.

While the traversal of a pulse is delay-insensitive allowing arbitrary wire delays, the downside is the timing assumption that the pulses should be long enough to

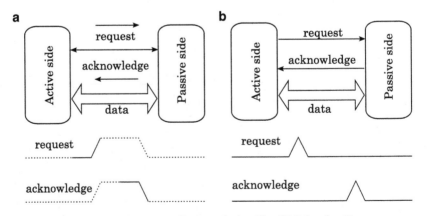

Fig. 4.15 More signalling schemes. (a) Single-track signalling (b) Pulse signalling

activate appropriate transitions at the receiving party. Additionally, it is required that the spaces between the pulses are long enough to be distinguished between consecutive pulses. These assumptions make it more difficult to implement over longer links and may have to be substituted with other coding techniques.

The pulse-mode technique was known for some time [22]. Initially, it only considered the non-conflicting events arriving and being processed by separate modules [41]. Later, it was also proposed for a more general CMOS-based implementation using arbiters [55]. Other design technologies such as RSFQ intrinsically favour such a pulse-based communication along with asynchronous design to ensure better performance [20].

Since pulse logic only uses wires during short non-overlapping time intervals, it is also possible to implement such communication over a single wire. Such a technique was demonstrated in the GasP communication protocol [55]. It was shown that while impressive results can be achieved, careful transistor-level design efforts are required.

4.7 SEU Resilient Codes

4.7.1 Phase Encoding

An additional source of problems exists in the wire interconnections: the possibility of transients occurring due to persistent noise such as cross-talk, cross-coupling or environmental interference. It is predicted that these *single event upsets* (SEUs) will have a greater rate in future deep sub-micron and nanometre technologies [36], and therefore techniques resilient to such a cause will become critical in future designs.

An approach addressing a degree of tolerance towards SEUs was proposed in [11]. The underlying principle is spreading the transmission event in time and space (between wires). The data are transported via the wires activated in a particular order. Each transaction requires both wires to make a transition. Due to this fact the channel is immune to the hazardous environment, which may produce glitches on individual wires. The system is only sensitive to glitches within the short intervals when a data symbol is transmitted.

Phase encoding is a self-synchronous protocol where the validity of data is sent together with the data. It is not speed-independent in the sense that it requires wires to preserve the order of transitions between transmitter and receiver. Figure 4.16 shows an example of two codewords being transmitted over two wires $t.0$ and $t.1$.

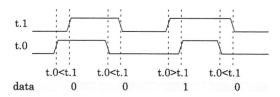

Fig. 4.16 Dual-rail phase encoding

Fig. 4.17 Phase encoding
repeater

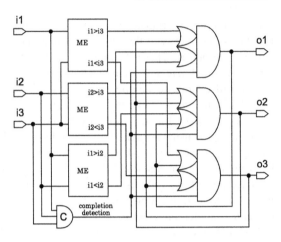

In the general case, N wires are able to encode $N!$ different codewords, which is
asymptotically greater than the dual-rail or M-of-N encoding [33].

The reception of a particular transition sequence for an N wired link is done
using a column of MUTEX (mutual exclusion or ME) elements. The ME elements
arbitrate between the transitions that occur on the pairs of wires and output a sequence of comparisons identifying the order of the incoming requests as a binary
relation. The completion detection is performed by a C-element function on all
wires. Figure 4.17 shows an implementation of a repeater. On the left side it detects a particular order of the events, while on the right side it reproduces the same
sequence whenever the completion detection signal is activated.

4.7.2 Data-Reference Codes

Another technique aimed at increasing the link resilience to Single Event signal
Upsets (SEUs) was proposed in [38]. It is based on redundancy and specific encoding. Each bit is represented by a pair of wires. One additional pair encodes the phase
reference that is used as a common guide for all other signals. Thus, the total number
of wires for the n-bit link is $(2 \cdot n) + 2$. Figure 4.18 demonstrates the functionality
of such a channel. The reference symbol always shifts bits according to the Johnson
code. The data wires transmit data based on the current value of the reference symbol. To transmit 0, it sends a copy of the reference symbol (in-phase transmission)
or the negated reference value for the transmission of 1.

Such an encoding has several interesting properties. First of all, it is able to detect when a single event upset occurs (Fig. 4.18). In all normal transitions, the data
symbol and reference symbol must always be either in phase or out of phase by
180°. In the case of an SEU, the Hamming distance between the signals becomes 1
(i.e. 90° out of phase). Second, transmitting one bit of data requires only one transition on the data wires and one transition of the reference symbol. Hence, the power
consumption is roughly similar to the NRZ dual-rail.

Fig. 4.18 Symbol and reference phase relationship [38]

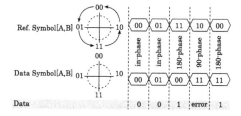

Table 4.2 Comparison of different links

Link	Wires	Bits encoded	Transitions/bit	Codewords	Is SI?	SEU resistant?
Dual-rail	2	1	2	2	Yes	No
1-of-4	4	2	1	4	Yes	No
1-of-4 LETS	4	2	0.5	4	Yes	No
LEDR	2	1	1	2	Yes	No
2-of-7	7	4	1	21	Yes	No
1-of-N	N	$\lfloor \log_2 N \rfloor$	$\frac{2}{\lceil \log_2 N \rceil}$	N	Yes	No
M-of-$2M$	$N = 2M$	$\lfloor \log_2 \binom{2M}{M} \rfloor$	$\frac{2M}{\lceil \log_2 \binom{2M}{M} \rceil}$	$\binom{2M}{M}$	Yes	No
Bundled data	$N = k+1$	$N-1$	$[2,\dots,2N]/(N-1)$	2^{N-1}	No	No
Phase-enc	N	$\lfloor \log_2 N! \rfloor$	$\frac{2N}{\lceil \log_2 N! \rceil}$	$N!$	No	Yes
Data-ref codes	$N = 2k+2$	$N/2 - 1$	N	$2^{N/2-1}$	Yes	Yes

4.7.3   Summary on Codes

As shown here, there is no universal encoding to serve all needs. Some are better in performance and reliability, while others in power, area or simplicity. Finding the right solution means finding the right balance and, possibly, tradeoffs between these qualities. Table 4.2 presents a short summary of the qualities of different encodings. The N value corresponds to the total number of wires used in the link and k is some natural number, $k > 0$.

The bundled data are good for its simplicity and scalability. It is similar to the codes used in synchronous design and is easy to use in computations. The number of transitions per bit can vary from $2/(N-1)$ (only the "data available" bit is changed) to $2N/(N-1)$ (all wires change their signal). Each additional wire effectively doubles the number of codewords that can be transmitted during one transaction. Its completion detection is based on the matched delay lines which reduces the flexibility of the design flow.

The main competitor of the bundled data is the dual-rail which is sometimes favoured by designers. Dual-rail uses two wires for each of the codeword bits as well as the additional circuitry to detect the completion of the signal; however, it is independent (see the "is SI?" column) of the delays on wires which is helpful for technology scaling. The completion detection circuitry does not necessarily make it slower than the bundled data, because the former code has to deal with variability and introduce timing margins for each of the delay lines. Additionally, the separate dual-rail lines are self-sufficient. Partial signal arrival can be used in *early evaluation* techniques [60] increasing overall system performance. For instance, the Boolean AND function can propagate the result without awaiting for the second argument, when the first one was 0.

The 1-of-4 (and similarly 1-of-N) encoding also provides the speed-independence and reduces the number of transitions per bit, which affects the consumption of power. For the increasing number of wires, the 1-of-N link quickly becomes infeasible to implement; however, multiple 1-of-4 codes can be combined to scale the bandwidth while providing a more efficient energy utilisation than it is with the dual-rail codes. Because of the spacers, all the 4-phase protocols have doubled the number of transitions which affects both the power and the throughput. Codes such as 1-of-4 LETS and LEDR alleviate the problem by using the 2-phase handshake, at the cost of additional circuit complexity.

The issue of choosing the right protocol is individual for each design. It is often the case that simple codes with area and latency overhead are preferred for short and numerous links, while the more sophisticated codes using lower wire transition count are used for communication over long wires. For instance, the Spinnaker system described in [42] uses the 4-phase 1-of-5 (Chain) links for the on-chip communication, while it resorts to the 2-phase 2-of-7 links for the inter-chip data flits.

Yet another issue associated with the link design is its resilience to signal distortion (see the "SEU resilience" column). As the risk of signal corruption increases, the role of the protocols that can detect and also correct errors in data is expected to be greater for the future chip interconnects.

4.8 Pipelining

With a reduced feature size, the RC delay for long interconnections grows exponentially [18]. As a result, communication on long wires is inevitably getting slower. To address such an issue, repeaters can be used to slow down the exponential growth of the wire delay by making it linear. However, for long links this still means a significant degradation of the link performance. This advocates the use of additional memory elements along the wires forming pipelines for the data traversal. Such pipelines would divide the critical path and increase the throughput of the circuit (Fig. 4.19). Traditional synchronous pipelines require a clock that causes additional consumption of power along with the increased complexity of routing the clock. Asynchronous pipelines in this respect may provide a convenient alternative [3]. Based on one of the handshake techniques shown before, they would be able to implement the handshake-based communication without great penalty in respect to performance, area and power consumption.

In this section, we consider the various ways of mitigating the performance penalty associated with the transmission of signals along handshake wires. For example, the use of explicit acknowledgements to separate data tokens adds a

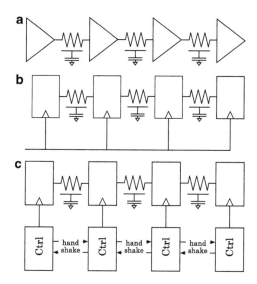

Fig. 4.19 Communication over long wires [16]. (**a**) Repeaters improve latency but do not help throughput (**b**) Synchronous latch pipelines improve latency and throughput but require a fast local clock (**c**) Asynchronous latch pipelines improve both latency and throughput and with low overhead

significant amount of delay to communication. One could look for ways where this cost may be hidden by concurrency, such as sensing the next request concurrently with the acknowledgement to the previous request. Other ways could be acking not every symbol but groups of symbols, or serialising to avoid large skew that has to be compensated for parallel interconnects.

4.8.1 Paired Handshake

As shown in [16], one way to deal with long wire communication would be to have paired control signals sharing the same data wires. This helps to hide the delay associated with the transmission of acknowledgement. Using the pulse-based GasP communication, the top and the bottom control signals take turns in sending the data over the shared channel (Fig. 4.20a). As soon as the top request signal arrives in the receiver and updates the corresponding memory latches, the bottom control pair may initiate its request without awaiting for the acknowledgement on the top side. Such a behaviour is depicted using Petri nets in Fig. 4.20b. Once the data from the first request was latched, the data channel may be reused by the bottom pair of handshakes. Therefore, the top and the bottom stage control logic is equipped with enabling signals *en_top* and *en_bot* mutually enabling each others logic at the right time.

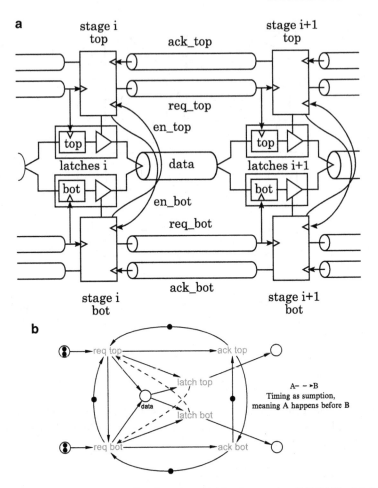

Fig. 4.20 Paired handshake increasing throughput. (**a**) Implementation [16] (**b**) Handshake Petri net description

Such a design was tested in 180-nm technology and it was reported to work slower for short links (less than 1.6 mm). However, for increasingly longer connections (1.6–5 mm), the performance benefit is getting close to the double of the original single-pair controlled transmission.

4.8.2 Serial vs. Parallel Links

Serial links as opposed to parallel links provide an alternative way to increase throughput over a long distance. The idea is to reuse the same wire to sequentially transmit several bits during the same transaction. Of course, such a transmission should be done at least N times faster than the transmission over corresponding

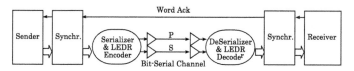

Fig. 4.21 Serial-Link structure [14]

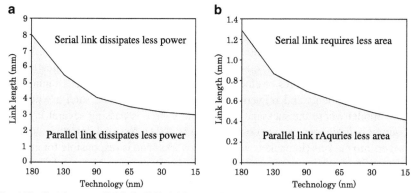

Fig. 4.22 Serial vs. parallel links [17]. (**a**) Power dissipation (**b**) Area

N-bit channel. To do that, one might think of skipping the acknowledgement phase for each separate bit and only wait for a single acknowledgement after a group of bits was transmitted. Such a group of bits is called the *word*. By transmitting words over the serial channel, fewer wires help to reduce the interconnect area, the number of line drivers, and the number of repeaters required. The circuit, however, requires additional logic to serialise and deserialise these words (Fig. 4.21). Also, special care needs to be taken to minimise crosstalk. It means that the area and the power consumption improvements from using such a technique only become obvious at a certain length of the interconnection. The length depends on the technology being used. As it can be seen from Fig. 4.22, as the technology is getting smaller, the serial link interconnect performs better for even shorter links. It makes the technique interesting for future long link and high throughput interconnects.

4.9   Networks-on-Chip

Since the problem of distributing a common clock with minimal skew is becoming more difficult and the design paradigm shifts towards the on-chip multi-processor architecture, planning multiple unrelated clock signals becomes a necessity. One convenient way to ensure the scaling is to provide an individual clock domain for each of the IP cores.

The systems built on the concept of multiple clocks are called the GALS (Globally Asynchronous, Locally Synchronous) systems. The GALS architecture was initially proposed in [8]. It allows to split the design into many independent modules

with individual clock rates and power supplies. Dividing the global clock routing
into smaller sub-tasks allows to reduce consumed power and simplify system design. Many smaller clock trees may potentially be run faster and increase system
performance; however, the latency introduced by communication may have a serious negative effect on the overall system performance [19].

Extensive studies have brought various gradations of GALS asynchrony. For
the low number of clocked domains, communication is built by directly interfacing these domains through special synchronous-to-synchronous (sync-sync) interfaces (Fig. 4.23). As the number of independent domains grows, a dedicated
communication medium called the *Network-on-Chip* is formed [1, 6, 37, 40, 46].

At the top level, the Network-on-Chip continues to serve as a communication link
connecting multiple IP cores or clock domains. However, it introduces a number of
new network elements and related concepts. We can associate such a NoC with
the OSI model where the message sent by the source is passing several layers of
abstraction before it reaches its destination (Fig. 4.24). Network interface converts
messages into packets (in packet switched networks) and is responsible for creating
and destroying links in circuit-switched networks. It also provides synchronization
between the associated clock domain and the network environment. The notice-
able standards for connecting IP cores to the network are: Open Core Protocol
(OCP) [39] and Advanced eXtensible Interface (AXI) [2].

At the network level, system has to provide a data propagation resource. It tries
to minimize the associated contention by choosing the appropriate topology and

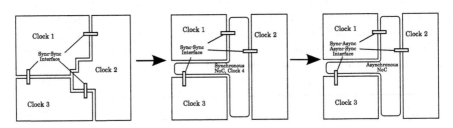

Fig. 4.23   GALS evolution

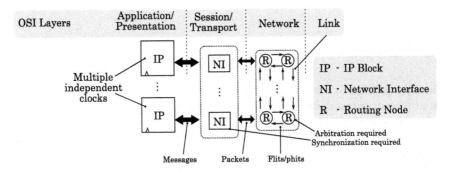

Fig. 4.24   Generic NoC building blocks

routing algorithm. One of the main challenges when building NoCs is ensuring quality of service. For instance, certain real-time applications may require the network to guarantee maximal propagation latency, minimal throughput or maximal power consumption. Simple *best effort* networks do not provide any guarantees; however, they are generically designed to provide the best quality of connection on average. The *guaranteed service* networks ensure the required specifications using additional connection resources. For instance, to guarantee the required latency, such networks might use additional communication links, use circuit switched routing or introduce prioritized communication channels. Alternatively, the artificial throughput bottleneck could effectively cap the power dissipation [40].

The NoC environment can use its own clock driving the communication components. Since this clock needs to cover the whole NoC area, the design has similar complications that are known in systems with the single global clock. Luckily, the number of links with significantly large clock phase shift is not overwhelming and special mesochronous synchronizers can be build on such links to address the issue [28].

The complete removal of the global clock from the NoC environment would be the next step of the evolution. The fully asynchronous environment for the global inter-chip communication eliminates the global clock distribution problem. The new challenges are the creation of the asynchronous routers and the asynchronous links.

4.10 Synchronizers

Once multiple independent clocks are used, the communication between different clock domains becomes an issue. Latches receiving data from a domain with a different clock may trigger reading at the moment when data are being written. Such a situation may result in non-digital behaviour of a latch which would become *metastable* (Fig. 4.25a). The metastable latch outputs a value that is between the logical 0 and 1 thresholds. Eventually it settles to 0 or 1; however, the metastability period might be long enough for the subsequent components to read the opposite value and cause unexpected circuit behaviour.

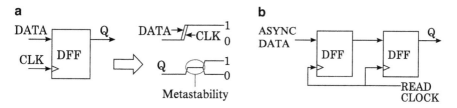

Fig. 4.25 Synchronization problem. (**a**) Metastability (**b**) Synchronizer

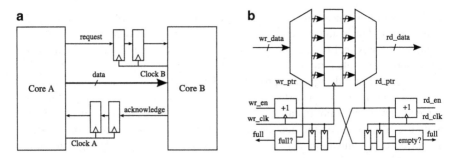

Fig. 4.26 Synchronizer solutions. (**a**) Synchronizers connecting clock domains (**b**) Bi-synchronous FIFO [60]

The reliable and efficient interfaces between synchronous islands is the key point in designing a GALS system [15]. If relative timing between the clocks is not possible to predict, simple synchronizers can be built by connecting two or more data flip-flops (Fig. 4.25b). The probability of the circuit malfunction can be calculated if the clock and the data change rates are known (see below).

Using simple two-flop synchronizers comes at a cost of additional synchronization latency and the reduced throughput (Fig. 4.26a). Note that between clocked domains the synchronizers are needed to do both: synchronize the data requests and the acknowledgements [24]. When the clocked network environment is used, this synchronizer area is doubled.

The combination of the overheads mentioned has made this circuit mostly of the theoretical interest. In practice, systems designed for small area and reduced power overhead use *pausable clocking* [34]. While the high performance oriented solutions use FIFOs, that add flexibility into the communication by decoupling the writer and the reader handshakes (Fig. 4.26b) [9, 32, 59].

4.10.1 Design of a Simple Synchronizer

Let us consider a design of a simple synchronizer, which receives an asynchronous data signal from the environment. This approach is also used to synchronize two clocked domains when their clocks are independent (in that case the frequency of the incoming data would be correlated with the second clock).

The asynchronous signal from the network may arrive at any moment regardless of what the state of the clock is (Fig. 4.25a). When both inputs of the DFF change at the same time, the output of the clocked latch component may become metastable and temporarily hold the value in the middle between logical 0 and 1. We assume here that the D-flip-flop is built of master and slave latches; so the master latch may enter metastability. Metastability may further propagate into slave on the back edge of the clock signal and be exposed outside the DFF. Such an output may be randomly interpreted by subsequent components and may cause circuit malfunction.

It is known that the timing of metastability is unbounded [24, 53]. However, the probability of the D-flip-flop being metastable P_{met} for a time t is as follows:

$$P_{\text{met}} = e^{-t/\tau} \tag{4.3}$$

where τ is a constant related to the technology used to build the DFF. Thus, this probability falls exponentially with the length of time. When two flip-flops are arranged in sequence as in Fig. 4.25b, the first flip-flop caught up in metastability will have an additional cycle to resolve it. Such a circuit would fail when the metastable state lasts longer than the period of one clock cycle and the unstable value is propagated to the second flip-flop. Depending on the technology, this probability may or may not be negligibly small. If such a probability is still large, more DFF components can be added. This way, by placing n DFF components, the probability of failure could be controlled:

$$P_{\text{met}\,n} = e^{-n \cdot t/\tau} \tag{4.4}$$

In different parts of the circuit, different n might be needed because the chance of the first DFF component becoming metastable depends on the frequency of the data arrival f_{data} and the clock frequency f_{clock}. Hence, if the flip-flop becomes metastable when both signals arrive within the Δ time interval, the chance that it goes metastable (during one second observation interval) can be written as:

$$P_{\text{go met}} = \Delta \cdot f_{\text{data}} \cdot f_{\text{clock}} \tag{4.5}$$

By combining (4.4) and (4.5), we can obtain the mean time between failures (MTBF):

$$\text{MTBF} = \frac{1}{P_{\text{go met}} \cdot P_{\text{met}\,n}} = \frac{e^{n \cdot t/\tau}}{\Delta \cdot f_{\text{data}} \cdot f_{\text{clock}}} \tag{4.6}$$

Typically, for most practical cases of system design $n = 2$ is sufficient [24].

Let us consider an example of MTBF calculation for a Jamb Latch synchronizer. For a $0.18\,\mu\text{m}$ technology and $Vdd = 1.5V$, deeper metastability has $\tau = 50ps$. Consider $\Delta = 10ps$, and $f_{data} = f_{clock} = 10^6 = 1\,GHz$. Suppose we wait for $t = 10\tau = 0.5ns$. Then

$$\text{MTBF} = \frac{e^{10}}{10^{-11} \cdot 10^{12}} = \frac{e^{10}}{10} \approx 2202.6 \approx 2 \cdot 10^3 \text{s}$$

which is less than 1 h.

Hence, for $MTBF = 10^7 - 10^8$ (100 days to 3 years), we need $t = 20\tau = 1ns$, which is on the borderline of $f_{clock} = 1GHz$. Hence, two DFFs are a justified measure.

4.11 Routers

The routers (also called switches) are the nodes connecting network links. Their basic function is to support different connectivity configurations and (for the packet switched networks) find the best propagation path depending on the routing algorithm used.

Based on the topology of the network, router would have certain number of input and output data channels, each associated with one input or output *port processor* (Fig. 4.27). These processors are used to buffer the incoming data and to convert the link encoding into the appropriate signals used withing the router.

The *switch controller* receives requests from the input port processors and accordingly configures the switching fabric module. Its main components are the *routing logic* making decisions about propagation direction and the *arbitration unit*. Since different propagation routes reuse the same links, it can happen that several input ports need to address the data to the same output port. In other words, the output port (and the link associated with it) becomes an exclusive resource requiring arbitration. The circuits resolving such conflicts in both synchronous and asynchronous designs are called *arbiters* (see below).

The *routing logic* is chosen based on the network topology and the network traffic patterns. Apart from finding the way to the destination core, its primary responsibility is to avoid deadlocks [12] stalling the system and livelocks that endlessly transport packets without finding the opportunity to deliver the information. The routing algorithms can be classified as the *deterministic routing* algorithms (when the route is uniquely identified by its source and destination coordinates) or the more complicated *adaptive routing* algorithms (which may take various routes depending on the congestion status of the network links).

The *switching fabric* (also called the *cross-bar*) connects its inputs and outputs in a point-to-point link fashion. Depending on the configuration received from the switch controller, it propagates its inputs to the outputs and may or may not allow concurrent connections of non-conflicting ports. Apart from the port buffers, the crossbar component is rather large and has complicated wiring. Its complexity grows quadratically with the number of inputs and outputs increased. Therefore, some routers reduce its size using *virtual channels*, *early ejection* and *look-ahead routing* that allow to split the large crossbar into two smaller ones [23].

Depending on the network, the router can be built either synchronous or asynchronous, thereby inheriting the advantages of each of the design approaches.

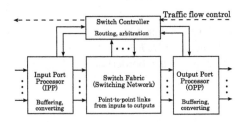

Fig. 4.27 Generic router structure [31]

4.11.1 Arbiters

While the synchronous arbiters are merely combinational functions, in the asynchronous environment such devices need to be able to deal with the requests arriving at any time and respond to the requests as they arrive.

The decision based on the relative time of the request arrival is difficult when this timing interval is less than the delay of the gates reacting to the requests. It may result in a metastable state when the circuit cannot decide which request to let through first. The problem of metastability in arbiters was noticed long time ago [25]. It was clear that depending on the timing interval in the arrival of the signals, the metastability can be arbitrarily long. An attempt was made to find the solution based on standard gates [43]; however, it was later realised that the circuit resolving metastability has to be analog. The first analog NMOS implementation was proposed in [47]. Later, such an element was presented in [29] for the CMOS technology (Fig. 4.28a) which we now know as the MUTEX (Mutual Exclusion) element. It is able to receive active requests on $r1$ and $r2$, and if it becomes metastable, both outputs would be held low until the metastability is resolved. Once designers had such a useful element, constructing more complicated arbitration units

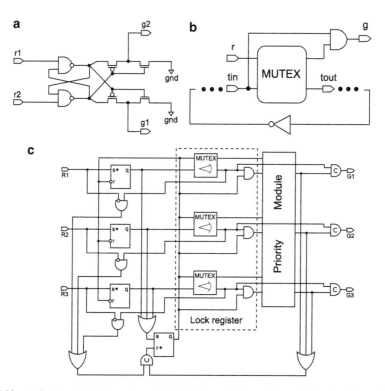

Fig. 4.28 Arbitration. (**a**) Mutex element [29] (**b**) Round-Robin arbiter (**c**) Priority arbiter [7]

was made possible [24]. A number of mutual exclusion elements could be combined in a tree, ring or a mesh to deal with multiple requests arriving at the same time (Fig. 4.28b).

Equipped with arbiters, asynchronous routers are able to manage conflicting requests and deal with the associated metastability. Depending on the application it is sometimes important to guarantee that the data are propagated from the sender to the receiver within a certain known interval of time. In asynchronous routers, it can be achieved using priority-based arbiters (Fig. 4.28c) [7]. Whenever the arbiter has choice among several active requests, it will favour the prioritised request first. This way the packets in high-priority channels can propagate within limited latency in highly congested networks.

4.12   CAD Issues

As opposed to the clocked design, where the clock signifies the updates of values in every bistable element in the system, the clock-less design allows partial system state updates. It creates a problem of considering all possible combinations of such states and makes the process much more complicated. Even small designs consisting of a dozen of gates may appear to be too complicated for manual design and require certain support of automated tools.

There is no standardised asynchronous design flow being widely used in industry [58]. A number of asynchronous tools, however, provide a reasonable help with designing circuits at a variable level of abstraction. Such tools can be roughly subdivided into the *logic synthesis* tools and the *syntax-driven design* tools [51].

4.12.1   Logic Synthesis

Logic synthesis is able to consider a system at the relatively low level of abstraction using such formalisms as Petri nets and, more specifically, signal transition graphs [10, 45]. STGs model the behaviour of a circuit by specifying possible sequences of transition events, similar to the way how Finite State Machines model possible sequences of system states in synchronous design. Unfortunately, these models easily become too complicated for large designs and in practice such an approach is only useful for the small-scale design.

Once the signal transition model is defined, it can be processed by synthesis tools such as Petrify [10]. Petrify traverses all reachable states, performs necessary checks for the implementability of the STG by a logic circuit (if necessary, solving the complete state assignment problem by inserting internal signals) and derives the gate-level implementation for each of the output and internal signals of the circuit. An example of designing control logic with STGs and Petrify will be presented at the end of this section.

For completeness, we also mention here another technique for logic synthesis of asynchronous controllers using Petri net and STG models. This technique is not so much restricted by the size of the specification, because it is based on the *direct mapping* of Petri net specifications into control logic, proposed in [62] and automated in [52]. This method was applied to the design of a duplex self-timed communication channel [67]. This channel's handshake protocol allows both master and slave to initiate the transmission of data from either side independently by means of a special protocol on two pairs of wires, which can be used at different phases of the protocol, for sending requests, acknowledgements and dual-rail data. An interested reader is recommended to refer to [67] for further details of the entire design process from the formal specification of the protocol to the circuit implementation of the entire link.

4.12.2   Syntax-Driven Design

Asynchronous design can be also based on translating some HDL program code to the circuit netlist. Examples of such techniques are TiDE (Haste) [58] and Balsa [5, 53]. Similar to VHDL or Verilog, Haste and Balsa specify the system behaviour. Such a program is then directly translated into a set of basic components which are then converted into a Verilog netlist suitable for further processing based on the traditional design flow. The approach provides a better option for building large systems. However, due to the limited set of basic components, it has a lower potential in designing an optimal system than the one based on STGs.

4.12.3   Example of Synthesis Using Petrify

Let us consider an example of designing a simple 4-phase controller for an asynchronous link interface based on the logic synthesis technique. It is a structure of the converter transceiving messages (packets) from the on-chip to the off-chip link (Fig. 4.29) proposed in [50]. Packets are formed of the data flits that arrive via the *internal data channel* pipeline into the Send Logic (SL) unit. To separate

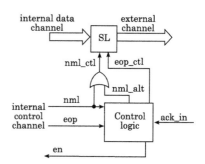

Fig. 4.29 Channel converter structure

different packets, the *internal control channel* provides additional information about the data flit types. When normal flit is sent, the *nml* signal is activated (*nml*+) by the control channel. Alternatively, when the last flit of the packet is sent, *eop* is activated (*eop*+).

The SL unit reacts to the incoming requests *nml_ctl*+ or *eop_ctl*+. For *nml_ctl*+, it propagates the flit from the internal data channel to the *external channel* link. For *eop_ctl*+, its task is to generate special "end-of-packet" flit and also send it across the off-chip link.

The Control Logic unit controls sending data over the external channel by activating either *nml_ctl*+ or *eop_ctl*+. Then, following the 4-phase protocol, it resets the request and acknowledges the control channel with *en*+. For the normal flit, after *nml*+ it only has to propagate the handshake: *nml*+ → *ack_in*+ → *en*+ and reset *nml*− → *ack_in*− → *en*−.

When the last flit activates *eop*+, the logic controller should, first, activate sending the flit via *nml_alt*+ → *nml_ctl*+ → *ack_in*+. Then, after the reset phase *nml_alt*− → *nml_ctl*− → *ack_in*−, it needs to send the "end-of-packet" flit and acknowledge the environment: *eop_ctl*+ → *ack_in*+ → *eop_ctl*− → *ack_in*−. The whole logic is modelled using the STG shown in Fig. 4.30a. We have a single token placed in the initial state $M0$ enabling two mutually exclusive transitions *eop*+ and *nml*+. It is now possible to follow a trace of events. For

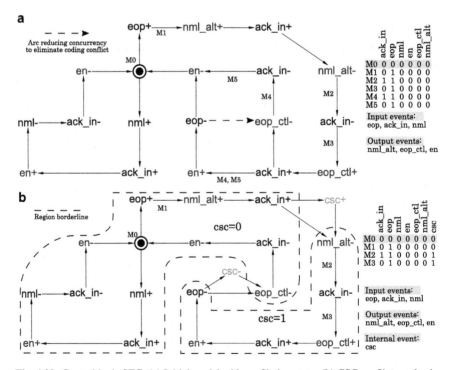

Fig. 4.30 Control logic STG. (**a**) Initial model with conflicting states (**b**) CSC conflict resolved

instance, when *eop*+ fires, the token will propagate to the arc *eop*+ → *nml_alt*+ which is denoted by marking $M1$. There is some concurrency in the diagram; after the arc *eop_ctl*+ → *ack_in*+ two concurrent tokens appear on the arcs *ack_in*+ → *en*+ and *ack_in*+ → *eon_ctl*−.

Implementing such an STG means finding the logic equation for each of the signals that the given circuit has to control. We can do that when the controlled signals are managed by the unique state codes. Such a property, however, does not hold in our initial diagram Fig. 4.30a. We can see, that, for instance, the state code for marking $M1$ is 010000, which is equal to that of $M3$. In other words, it means that while being in such a state the circuit logic cannot decide which event should occur next: *nml_alt*+ or *eop_ctl*+. There are more such conflicts, $M2$ and $M4$ or $M1$ and $M5$. To deal with the problem, and achieve the complete state coding, we can constraint the model by adding new arcs (adding more conditions for some event to fire). In order not to change the model of the environment behaviour, we can only add such constraints for the internal or the output events. For instance, if we create an arc *eop*− → *eop_ctl*− (Fig. 4.30a), the event *eop_ctl*− will never occur before *eop*−. It will reduce concurrency and markings $M4$ with $M5$ will become unreachable. This allows us to sort out the state coding for $M2$. $M5$ is also eliminated; however, $M1$ is still in conflict with $M3$. To resolve this conflict, we can increase the number of signals we use to encode the states. For instance, if we add a new sequential signal *csc* as shown in Fig. 4.30b, it will split the diagram into two regions, one with *csc* = 0 and the other with *csc* = 1. When doing so, we only need to achieve that the conflicting markings appear in different regions. In our case the new unique coding for $M1$ and $M3$ are: $M1 = 0100000$ and $M3 = 0100001$.

By giving this model to Petrify, it is possible to find the implementation for each of the signals, which is:

$$en = ack\_in \cdot (\overline{eop} + eop\_ctl)$$
$$eop\_ctl = csc \cdot (eop\_ctl + \overline{ack\_in})$$
$$nml\_alt = \overline{csc} \cdot eop$$
$$csc = eop \cdot (csc + ack\_in)$$

The process of adding new signals and reducing concurrency can also be done automatically with Petrify. One may find that there are also other solutions resolving the conflicts without sacrificing the concurrency; in our case it could be done by adding two additional signals instead of one. Whenever we have a choice as to which approach to use, there are certain tradeoffs related, for example, to the performance of the overall system, including both the controller and the controlled logic. Concurrency reduction makes the controller implementation simpler, due to having more don't cares in the circuit state space, but may sometimes either degrade the system's performance, depending on the mean value and distribution of the delays associated with the initially concurrent branches. On the other hand, the insertion of additional signals implies a more complex controller implementation, which may reduce the performance if the gains from the concurrent branches in the controlled circuit are relatively small. In the latter case, basically the gains are completely absorbed by the delays introduced by the cells that implement csc signals.

Yet another powerful technique for resolving state encoding conflicts is based on applying timing assumptions. It can greatly simplify the implementation without the additional performance penalty. It is similar to concurrency reduction; however, the events are not specifically constrained by adding new arcs. We only state that certain events happen in a predefined order. For instance, we know that the external link channel is slow, much slower than any gate in the control logic. This allows us to model some of the events as sequential, i.e. $nml+ \rightarrow ack\_in+$ and $nml- \rightarrow ack\_in-$. This, in turn, simplifies the implementation for the signal en because it does not have to sense the nml signal at all.

4.13 Conclusions

The chapter presents currently known link level techniques while addressing the challenges of modern on-chip communication. It mentions the key concepts of the communication by relating them to the token-based view of a communication link. Starting from the basics of asynchronous communications such as signalling techniques, handshake phases and channel types, the chapter advances into the overview of different encoding styles and mentions a variety of the associated benefits such as simplicity, performance, power consumption or the resilience towards single event transients. On the higher level of the communication, basics of the NoC structure are noted. In particular, it mentions basic NoC components. It identifies and explains the problem of metastability which manifests itself in the asynchronous world (arbiters) and on the borderline with the clocked domains (synchronizers). Finally, the chapter provides a glance at some of the existing asynchronous design tools. It gives references to more sources for an interested reader and finalises the chapter with a more detailed example of designing a simple circuit controller based on a given STG model.

It would be impossible to cover all the aspects of asynchronous communication in a single review; however, it was shown that the asynchronous design provides a great variety of unexplored directions and has a fertile ground for further research.

Acknowledgements This work was partially supported by EPSRC EP/E044662/1. We are grateful to Robin Emery for his useful comments on this chapter.

References

1. Agarwal, A., Iskander, C., Shankar, R.: Survey of network on chip (NoC) Architectures & Contributions. J. Eng. Comput. Archit. **3**(1) (2009)
2. AMBA Advanced eXtensible Interface (AXI) protocol specification, version 2.0 www.arm.com
3. Bainbridge, J., Furber, S.: CHAIN: A delay-insensitive chip area interconnect. IEEE Micro **22**, 16–23 (2002)

4. Bainbridge, W.J., Toms, W.B., Edwards, D.A., Furber, S.B.: Delay-insensitive, point-to-point interconnect using m-of-n codes. In: Proc. Ninth International Symposium on Asynchronous Circuits and Systems, pp. 132–140 (2003)
5. Bardsley, A.: Implementing Balsa handshake circuits. Ph.D. thesis, Department of Computer Science, University of Manchester (2000)
6. Bjerregaard, T., Mahadevan, S.: A survey of research and practices of Network-on-Chip. ACM Comput. Surv. 38(1), 1 (2006)
7. Bystrov, A., Kinniment, D.J., Yakovlev, A.: Priority arbiters. In: ASYNC '00: Proceedings of the 6th International Symposium on Advanced Research in Asynchronous Circuits and Systems, pp. 128–137. IEEE Computer Society, Washington, DC, USA (2000)
8. Chapiro, D.: Globally asynchronous locally synchronous systems. Ph.D. thesis, Stanford University (1984)
9. Chelcea, T., Nowick, S.M.: Robust interfaces for mixed-timing systems. IEEE Trans. VLSI Syst. 12(8), 857–873 (2004)
10. Cortadella, J., Kishinevsky, M., Kondratyev, A., Lavagno, L., Yakovlev, A.: Logic synthesis of asynchronous controllers and interfaces. Springer, Berlin, ISBN: 3-540-43152-7 (2002)
11. D'Alessandro, C., Shang, D., Bystrov, A.V., Yakovlev, A., Maevsky, O.V.: Multiple-rail phase-encoding for NoC. In: Proc. International Symposium on Advanced Research in Asynchronous Circuits and Systems, pp. 107–116 (2006)
12. Dally, W.J., Seitz, C.L.: Deadlock free message routing in multiprocessor interconnection networks. IEEE Trans Comput C-36(5), 547–553 (1987)
13. Dean, M., Williams, T., Dill, D.: Efficient self-timing with level-encoded 2-phase dual-rail (LEDR). In: C.H. Séquin (ed.) Advanced Research in VLSI, pp. 55–70. MIT Press, Cambridge (1991)
14. Dobkin, R., Perelman, Y., Liran, T., Ginosar, R., Kolodny, A.: High rate wave-pipelined asynchronous on-chip bit-serial data link. In: Asynchronous Circuits and Systems, 2007. ASYNC 2007. 13th IEEE International Symposium on, pp. 3–14 (2007)
15. Ginosar, R.: Fourteen ways to fool your synchronizer. In: Proc. International Symposium on Advanced Research in Asynchronous Circuits and Systems, pp. 89–96. IEEE Computer Society Press, Washington, DC (2003)
16. Ho, R., Gainsley, J., Drost, R.: Long wires and asynchronous control. In: Proc. International Symposium on Advanced Research in Asynchronous Circuits and Systems, pp. 240–249. IEEE Computer Society Press, Washington, DC (2004)
17. Ho, R., Horowitz, M.: Lecture 9: More about wires and wire models. Computer Systems Laboratory (2007)
18. International technology roadmap for semiconductors: 2005 edition www.itrs.net/Links/2005ITRS/Home2005.htm. Cited 10 Nov 2009
19. Iyer, A., Marculescu, D.: Power and performance evaluation of globally asynchronous locally synchronous processors. In: ISCA '02: Proceedings of the 29th annual international symposium on Computer architecture, pp. 158–168. IEEE Computer Society, Washington, DC, USA (2002)
20. Kameda, Y., Polonsky, S., Maezawa, M., Nanya, T.: Primitive-level pipelining method on delay-insensitive model for RSFQ pulse-driven logic. In: Proceedings of the Fourth International Symposium on Advanced Research in Asynchronous Circuits and Systems, pp. 262–273 (1998)
21. Käslin, H.: Digital Integrated Circuit Design: From VLSI Architectures to CMOS Fabrication. Cambridge University Press, Cambridge, ISBN: 978-0-521-88267-5 (2008)
22. Keller, R.: Towards a theory of universal speed-independent modules. IEEE Trans Comput C-23(1), 21–33 (1974)
23. Kim, J., Nicopoulos, C., Park, D.: A gracefully degrading and energy-efficient modular router architecture for on-chip networks. SIGARCH Comput. Archit. News 34(2), 4–15 (2006)
24. Kinniment, D.J.: Synchronization and Arbitration in Digital Systems. John Wiley & Sons, Ltd (2007)
25. Kinniment, D.J., Edwards, D.: Circuit technology in a large computer system. In: Proceedings of the Conference on Computers–Systems and Technology, pp. 441–450 (1972)

26. Lin, T., Chong, K.S., Gwee, B.H., Chang, J.S.: Fine-grained power gating for leakage and short-circuit power reduction by using asynchronous-logic. In: IEEE International Symposium on Circuits and Systems, 2009 (ISCAS 2009), pp. 3162–3165 (2009)
27. Liu, Y., Nassif, S., Pileggi, L., Strojwas, A.: Impact of interconnect variations on the clock skew of a gigahertz microprocessor. In: Proceedings of the 37th Conference on Design Automation, 2000. Los Angeles, CA, pp. 168–171 (2000)
28. Ludovici, D., Strano, A., Bertozzi, D., Benini, L., Gaydadjiev, G.N.: Comparing tightly and loosely coupled mesochronous synchronizers in a NoC switch architecture. In: 3rd ACM/IEEE International Symposium on Networks on Chip, pp. 244 – 249 (2009)
29. Martin, A.J.: Programming in VLSI: From communicating processes to delay-insensitive circuits. In: C.A.R. Hoare (ed.) Developments in Concurrency and Communication, UT Year of Programming Series, pp. 1–64. Addison-Wesley (1990)
30. McGee, P., Agyekum, M., Mohamed, M., Nowick, S.: A level-encoded transition signaling protocol for high-throughput asynchronous global communication. In: 14th IEEE International Symposium on Asynchronous Circuits and Systems, 2008 (ASYNC '08), pp. 116–127 (2008)
31. Mir, N.F.: Computer and Communication Networks. Prentice Hall, Englewood Cliffs, NJ (2006)
32. Miro Panades, I., Greiner, A.: Bi-synchronous fifo for synchronous circuit communication well suited for network-on-chip in gals architectures. In: NOCS '07: Proceedings of the First International Symposium on Networks-on-Chip, pp. 83–94. IEEE Computer Society, Washington, DC, USA (2007)
33. Mokhov, A., D'Alessandro, C., Yakovlev, A.: Synthesis of multiple rail phase encoding circuits. In: 15th IEEE Symposium on Asynchronous Circuits and Systems, 2009 (ASYNC '09), pp. 95–104 (2009)
34. Mullins, R., Moore, S.: Demystifying data-driven and pausible clocking schemes. In: ASYNC '07: Proceedings of the 13th IEEE International Symposium on Asynchronous Circuits and Systems, pp. 175 185. IEEE Computer Society, Washington, DC, USA (2007)
35. Murata, T.: Petri nets: Properties, analysis and applications. In: Proceedings of the IEEE, vol. 77, pp. 541–580 (1989)
36. Nicolaidis, M.: Time redundancy based soft-error tolerance to rescue nanometer technologies. In: Proceedings of the 17th IEEE VLSI Test Symposium, 1999, pp. 86–94 (1999)
37. Nicopoulos, C., Narayanan, V., Das, C.R.: Network-on-Chip Architectures. Springer, Berlin (2009)
38. Ogg, S., Al-Hashimi, B., Yakovlev, A.: Asynchronous transient resilient links for NoC. In: CODES/ISSS '08: Proceedings of the 6th IEEE/ACM/IFIP international conference on Hardware/Software codesign and system synthesis, pp. 209–214. ACM, New York, NY, USA (2008)
39. Open Core Protocol www.ocpip.org.
40. Pande, P.P., Grecu, C., Jones, M., Ivanov, A., Saleh, R.: Performance evaluation and design trade-offs for network-on-chip interconnect architectures. IEEE Trans Comput **54**(8), 1025–1040 (2005)
41. Plana, L., Unger, S.: Pulse-mode macromodular systems. In: Proceedings of the International Conference on Computer Design: VLSI in Computers and Processors, 1998 (ICCD '98), pp. 348–353 (1998)
42. Plana, L.A., Furber, S.B., Temple, S., Khan, M., Shi, Y., Wu, J., Yang, S.: A gals infrastructure for a massively parallel multiprocessor. IEEE Des. Test **24**(5), 454–463 (2007)
43. Plummer, W.: Asynchronous arbiters. IEEE Trans Comput **C-21**(1), 37–42 (1972)
44. Poliakov, I., Khomenko, V., Yakovlev, A.: Workcraft — a framework for interpreted graph models. In: PETRI NETS'09: Proceedings of the 30th International Conference on Applications and Theory of Petri Nets, pp. 333–342. Springer-Verlag, Berlin, Heidelberg (2009)
45. Rosenblum, L., Yakovlev, A.: Signal graphs: from self-timed to timed ones. Int. Workshop on Timed Petri Nets, pp. 199–206 (1985)
46. Salminen, E., Kulmala, A., Hamalainen, T.D.: Survey of network-on-chip proposals. www.ocpip.org/white_papers.php. (2008)
47. Seitz, C.L.: Ideas about arbiters. Lambda **1**, 10–14 (1980)

48. Shams, M., Ebergen, J.C., Elmasry, M.I.: Modeling and comparing CMOS implementations of the C-element. IEEE Trans. VLSI Syst. **6**(4), 563–567 (1998)
49. Shang, D., Yakovlev, A., Koelmans, A., Sokolov, D., Bystrov, A.: Dual-rail with alternating-spacer security latch design.   Tech. Rep. NCL-EECE-MSD-TR-2005-107, Newcastle University (2005)
50. Shi, Y., Furber, S., Garside, J., Plana, L.: Fault tolerant delay-insensitive inter-chip communication. In: 15th IEEE Symposium on Asynchronous Circuits and Systems, 2009 (ASYNC '09), pp. 77–84 (2009)
51. Sokolov, D., Yakovlev, A.: Clockless circuits and system synthesis. IEE Proc. Digit. Tech. **152**(3), 298–316 (2005)
52. Sokolov, D., Bystrov, A., Yakovlev, A.: Direct mapping of low-latency asynchronous controllers from stgs. IEEE Transactions on Computer-Aided Design of Integrated Circuits and Systems **26**(6), 993–1009 (2007)
53. Sparsø, J., Furber, S.: Principles of Asynchronous Circuit Design. Kluwer Academic Publishers, ISBN: 978-0-7923-7613-2, Boston/Dordrecht/London (2002)
54. Sutherland, I.E.: Micropipelines. Commun ACM **32**(6), 720–738 (1989)
55. Sutherland, I.E., Fairbanks, S.: GasP: A minimal FIFO control. In: Proc. International Symposium on Advanced Research in Asynchronous Circuits and Systems, pp. 46–53. IEEE Computer Society Press (2001)
56. Sutherland, I.E., Sproull, R.: The theory of logical effort: Designing for speed on the back of an envelope. In: Proc. IEEE Advanced Research in VLSI, pp. 3–16. UC Santa Cruz (1991)
57. Sutherland, I.E., Molnar, C.E., Sproull, R.F., Mudge, J.C.: The Trimosbus. In: C.L. Seitz (ed.) Proceedings of the First Caltech Conference on Very Large Scale Integration, pp. 395–427 (1979)
58. Taubin, A., Cortadella, J., Lavagno, L., Kondratyev, A., Peeters, A.M.G.: Design automation of real-life asynchronous devices and systems. Foundations and Trends in Electronic Design Automation **2**(1), 1–133 (2007)
59. Thonnart, Y., Beigne, E., Vivet, P.: Design and implementation of a gals adapter for anoc based architectures. In: 15th IEEE Symposium on Asynchronous Circuits and Systems, 2009 (ASYNC '09), pp. 13–22 (2009)
60. Thornton, M.A., Fazel, K., Reese, R.B., Traver, C.: Generalized early evaluation in self-timed circuits. In: Proc. Design, Automation and Test in Europe (DATE), pp. 255–259 (2002)
61. van Berkel, K., Bink, A.: Single-track handshake signaling with application to micropipelines and handshake circuits. In: Advanced Research in Asynchronous Circuits and Systems, 1996. Proceedings., Second International Symposium on, pp. 122–133 (1996)
62. Varshavsky, V.I., Kishinevsky, M.A., Marakhovsky, V., Yakovlev, A.: Self-Timed Control of Concurrent Processes: The Design of Aperiodic Logical Circuits in Computers and Discrete Systems. Kluwer Academic Publishers, Dordrecht, The Netherlands (1990)
63. Varshavsky, V., Marakhovsky, V., Rosenblum, L., Tatarinov, Y.S., Yakovlev, A.: Towards fault tolerant hardware implementation of physical layer network protocols. In: Automatic Control and Computer Science (translated from Russian), vol. 20, pp. 71–76. Allerton Press (1986)
64. Varshavsky, V.I., Volodarsky, V.Y., Marakhovsky, V.B., Rozenblyum, L.Y., Tatarinov, Y.S., Yakovlev, A.: Structural organization and information interchange protocols for a fault-tolerant self-synchronous ring baseband channel (pt.1). Hardware implementation of protocols for a fault-tolerant self-synchronous ring channel (pt.2). Algorithmic and structural organization of test and recovery facilities in a self-synchronous ring (pt.3), vol. 22, no 4, pp. 44 – 51 (pt.1), no 5, pp. 59 – 67 (pt.2), vol. 23, no 1, pp. 53 – 58 (pt.3). In: Automatic Control and Computer Science. Allerton Press, Inc. (1988, 1989)
65. Verhoeff, T.: Delay-insensitive codes—an overview. Distr Comput **3**(1), 1–8 (1988)
66. Visweswariah, C.: Death, taxes and failing chips. In: DAC '03: Proceedings of the 40th annual Design Automation Conference, pp. 343–347. ACM, New York, NY, USA (2003)
67. Yakovlev, A., Furber, S., Krenz, R., Bystrov, A.: Design and analysis os a self-timed duplex communication system. IEEE Trans Comput **53**(7), 798–814 (2004)
68. Yakovlev, A., Varshavsky, V., Marakhovsky, V., Semenov, A.: Designing an asynchronous pipeline token ring interface. In: Asynchronous Design Methodologies, pp. 32–41. IEEE Computer Society Press (1995)

Part II
System-Level Design Techniques

Chapter 5
Application-Specific Routing Algorithms for Low Power Network on Chip Design

Maurizio Palesi, Rickard Holsmark, Shashi Kumar, and Vincenzo Catania

Abstract In the last few years, Network-on-Chip (NoC) has emerged as a dominant paradigm for synthesis of a multi-core Systems-on-Chip (SoC). A future NoC architecture must be general enough to allow volume production and must have features for specialization and configuration to match and meet the application's power and performance requirements. This chapter describes how one important aspect, namely the routing algorithm, can be optimized in such NoC platforms. Routing algorithm has a major effect on the performance (packet latency and throughput) as well as power consumption of NoC. A methodology to develop efficient and deadlock free routing algorithms which are specialized for an application or a set of concurrent applications is presented. The methodology, called application-specific routing algorithms (APSRA), exploits the application-specific information regarding pairs of cores which communicate and other pairs which never communicate in the NoC platform. This information is used to maximize the adaptivity of the routing algorithm without compromising the important property of deadlock freedom. The chapter also presents an extensive comparison between the routing algorithms generated using APSRA methodology and general purpose deadlock-free routing algorithms. The simulation-based evaluations are performed using both synthetic traffic and traffic from real applications. The comparison embraces several performance indices such as degree of adaptiveness, average delay, throughput, power dissipation, and energy consumption. In spite of an adverse impact on router architectural complexity, it is shown that the higher adaptivity of APSRA leads to significant improvements in both routing performance and energy consumption.

5.1 Introduction

In the last few years Network-on-Chip (NoC) has emerged as a dominant paradigm for synthesis of multi-core Systems-on-Chip (SoC). The main question has now shifted from "whether to use or not to use NoC" to "how to effectively use NoC

M. Palesi (✉)
Dipartimento di Ingegneria Informatica e delle Telecomunicazioni, University of Catania, Italy
e-mail: mpalesi@diit.unict.it

C. Silvano et al. (eds.), *Low Power Networks-on-Chip*,
DOI 10.1007/978-1-4419-6911-8_5, © Springer Science+Business Media, LLC 2011

paradigm in practice." A future NoC architecture must be general enough to allow volume production and must have features for specialization and configuration to match and meet the applications performance requirements. It is possible to envision a chip based on NoC paradigm as a future FPGA. Such a field programmable resource array (FPRA) will use grosser level of configurable computational resources and grosser level of configurable communication resources for packet-switched communication among on-chip system components. Already there have been proposals in this direction of designing programmable NoC platforms [17, 19, 45]. One can easily envision a scenario in which a mesh topology NoC chip, populated with an application area-specific set of cores, could be available as an off-the-shelf standard product. One can easily imagine that one such multi-core chip will be next in line to the current superscalar DSP chips for building multi-media gadgets. Lowering the power consumption of these platform chips will be important since most of the gadgets built using them will be battery operated.

This chapter describes one way of specialization of such programmable NoC platforms. Routing algorithm has a major effect on the performance (packet latency and throughput) as well as power dissipation and energy consumption of NoC. An adaptive routing algorithm with low average latency and high average throughput is also likely to be efficient for power dissipation and energy consumption. This chapter describes a methodology to develop routing algorithms which are both performance and power efficient as well as are deadlock free. These routing algorithms are specialized for an application or a set of concurrent applications. The methodology is called application specific routing algorithms (APSRA). APSRA methodology is based on the realization that in the embedded systems, unlike general networks, the computational as well as communication requirements of the applications can be very well characterized. Specifically, it is known which pairs of cores in the NoC system communicate and which do not. By off-line profiling and analysis, one can also estimate some quantitative information like communication bandwidth requirements between communicating pairs. After the applications have been mapped and scheduled on the NoC platform, information about communications which are never concurrent is also available. These communication characteristics and requirements of the application can be used to optimize network design cost and its operating performance. One research group has used information about the application to design an application-specific NoC topology [6] which is most suitable to the application and will lead to the best performance for it. Guz et al. [20] has proposed methods to specialize link bandwidth of a mesh NoC in order to match them with the communication requirements of a mapped application. They demonstrate that their specialization leads to significant reduction in power consumption. It is obvious that there is a large scope for using information regarding the application to effectively design as well as use a NoC platform. These pieces of application-specific information are used in this chapter to optimize the routing algorithm for NoC.

The goal of this chapter is to demonstrate how network routing performance (and indirectly its energy consumption) can be improved by using application communication characteristics. A comprehensive survey of routing algorithms for

NoC platforms highlighting the implications on router cost, power and timing is also given. An application-specific routing algorithm is intuitively expected to improve communication performance as well as energy consumption. The chapter will also present an extensive comparison between the routing algorithms generated using APSRA methodology with general purpose deadlock-free routing algorithms. The evaluations will analytically assess the adaptivity provided by the routing algorithms, whereas network simulations, with both synthetic an real traffic scenarios, will be used to estimate the dynamic routing performance. The comparison embraces several performance indices such as degree of adaptiveness, average delay, throughput, power dissipation and energy consumption. Implications of APSRA on router architecture, area, and power dissipation are also discussed. It will be shown that increase in performance comes at the cost of area and power consuming tables in the router. Thus, a method to compress the routing table to reduce this overhead and disadvantage of APSRA is presented. In the end, the importance of APSRA methodology in specializing general purpose NoC platforms for optimizing application's communication performance and energy consumption is summarized.

The chapter is organized as follows. Section 5.2 provides some background and related work in the area of routing algorithms for NoCs. Section 5.3 defines some terminology and provides the theoretical basis for APSRA methodology. The APSRA design methodology is presented in Sect. 5.4. Sections 5.5 and 5.6 present performance evaluation and power analysis of APSRA, respectively. Finally, Sect. 5.7 summarizes the chapter and draws directions for future developments.

5.2 Background on Routing Algorithms and Power Dissipation

The overall performance of a network depends on several network properties such as topology, routing algorithm, flow control and switching technique. This chapter mainly focuses on routing algorithms. A routing algorithm determines the path selected by a packet to reach its destination. This section provides a brief overview on routing algorithms focusing on wormhole switching technique and deadlock issues. Then, it discusses about the two main components of any routing algorithm namely the routing function and the selecion function. Hardware implications due to routing algorithm will be discussed and its impact on power metrics will be qualitatively analyzed. Finally, common performance metrics used to evaluate routing algorithms will be presented.

5.2.1 Classification of Routing Algorithms

Routing schemes have been classified in several ways in literature. In a scheme called *source routing*, the source node selects the entire path before sending the packet. The major drawback of this approach is that each packet must carry this path information, thus increasing the packet size. In addition, the path cannot be changed after the packet has left the source. A more common solution is the use of *distributed routing*. Here, a router upon receiving a packet decides whether it should be delivered to the local resource or forwarded to a neighboring router. In the latter case, a routing algorithm is invoked (or a routing table is accessed) to determine which neighbor the packet should be sent.

Routing algorithms can also be classified as *deterministic*, *oblivious*, and *adaptive*. In deterministic routing, the path from the source to the destination is completely determined by the source and the destination addresses. In both oblivious and adaptive routing, multiple paths from the source to the destination are possible. The difference between them is the way in which a given path is selected. In oblivious routing, the selection of the routing path does not depend on the current status of the network, whereas in adaptive routing the selection of the routing path is determined by the current status of the network (e.g., congested channels and hot-spot regions).

5.2.2 *Wormhole Switching and Deadlock*

Differently from virtual cut-through and circuit switching commonly used in data networks, *wormhole* switching [11] does not require large packet buffers at each intermediate node, but only a small FIFO flit buffer is required. This is one of the most important points in favour of using wormhole switching in NoC as buffers represent the main contribution in cost (silicon area) and power dissipation of the router (c.f. Sect. 5.6). In addition, the pipeline nature of wormhole makes the network latency relatively insensitive to path length.

Unfortunately, wormhole routing is particularly susceptible to *deadlocks*. Deadlock is the situation in which a set of packets is blocked forever in a circular wait in the network. Such an event happens because packets are allowed to hold some resources while requesting others. In the case of wormhole switching, resources are channels (and their corresponding flit buffers).

Figure 5.1 shows an example of deadlock situation [35] occurring when, at the same time t, four packets *Packet 1*, *Packet 2*, *Packet 3*, and *Packet 4*, present at west port of router 3, south port of router 4, east port of router 2, and north port of router 1 respectively. Destinations of the packets are two hops counterclockwise: Packet 1 is destined to node 2, Packet 2 to node 1, Packet 3 to node 3, and Packet 4 to node 4. Assuming a routing function which returns the output port to reach the destination in a minimal hop count favouring counterclockwise direction, at time $t + 1$ the input logic of router 3 selects the west input creating a shortcut between the west input port and the east output port. Based on wormhole rules, such a shortcut is mantained until all the flits of Packet 1 have traversed router 3. At the same time, a shortcut from south input port of router 4 with its north output port is created. Similarly, a shortcut between east input port to west output port, and from north input port to south output port are created at time $t + 1$ in router 2 and router 1, respectively. At time $t + 2$, the first flit of Packet 1 is stored into the west input buffer of router 4.

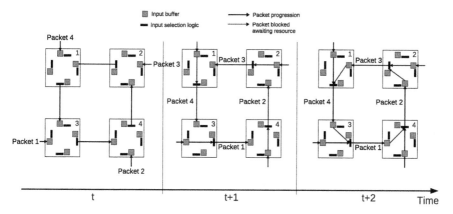

Fig. 5.1 An example of deadlock

The routing function determines that the flit has to be forwarded to the north output port, but this port is already assigned for forwarding the flits of Packet 2. Thus, the first flit of Packet 1 is blocked in the west input buffer of router 4. Similarly, the first flit of Packet 2 is blocked in the south input port of router 2, the first flit of Packet 3 is blocked in the east input port of router 1, and the first flit of Packet 4 is blocked in the north input port of router 3. Assuming 1 flit input buffer size, the flits of the four packets cannot advance and, consequently, there is a deadlock.

5.2.3 Basic Components of a Routing Algorithm

In general, a routing algorithm can be seen as the cascade of two main blocks that implement the routing function and a selection function (a.k.a., selection policy or selection strategy), respectively, as shown in Fig. 5.2. First, a *routing function* computes the set of admissible output channels towards which the packet can be forwarded to reach the destination. Then, a *selection function* is used to select one output channel from the set of admissible output channels returned by the routing function. Of course, in a router implementing a deterministic routing algorithm, the selection block is not present since the routing function returns only a single output port (Fig. 5.2a). In a router implementing an oblivious routing algorithm the selection block takes its decision based solely on the information provided by the header flit (Fig. 5.2b). Finally, network status information (e.g., link utilization and buffer occupation) are exploited by the selection function of a router implementing an adaptive routing algorithm (Fig. 5.2c).

5.2.3.1 Routing Function

Many routing functions for wormhole-switched networks have been proposed in the literature [9, 10, 18, 46]. Glass and Ni in [18] propose a turn-model for designing

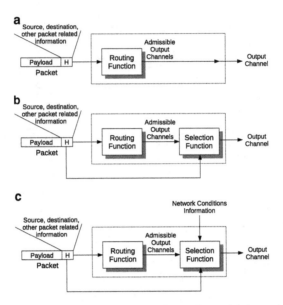

Fig. 5.2 Routing and selection blocks of a router for (**a**) Deterministic routing (**b**) Oblivious routing (**c**) Adaptive routing

wormhole deadlock free routing algorithms for mesh and hypercube topology networks. Prohibiting enough turns to avoid formation of any cycle produces routing algorithms that are deadlock free, livelock free, and highly adaptive. This model has been later utilized by Chiu in [10] to develop the *Odd-Even* adaptive routing algorithm for meshes without virtual channels. The model restricts the locations where some turns can be taken so that deadlock is avoided. In comparison with the turn-model, the degree of routing adaptiveness provided by the model is more even for different source-destination pairs. A non-minimal deadlock-free routing algorithm is described for an irregular mesh topology NoC with regions in [22]. This algorithm is biased in favor of some areas of the network as compared to the other areas.

Many other routing algorithms have been proposed in the literature, especially in the domain of high-performance computing based on a cluster of PCs. Some of them provide fault tolerance by the use of topology-agnostic routing, possibly combined with a reconfiguration process. A few example of algorithms that do not require virtual channels and that can be adapted to be used in the NoC domain are L-turn [26], smart-routing [8], FX [42], and segment-based routing [32].

5.2.3.2 Selection Function

The potentiality of a routing function can be exploited if it is coupled with an effective selection function. In fact, even the most flexible routing function, able to provide many routing alternatives to reach the destinations, could perform poorly if it is not coupled with an appropriate selection function. Several selection strategies aimed at exploiting the potential of highly adaptive routing algorithms have been proposed in literature. Hu and Marculescu in [24] propose a routing scheme called *DyAD* which combines the advantages of both deterministic and adaptive routing schemes. The router works in a deterministic mode when the network is not congested, and switches to an adaptive mode when the network becomes congested. In [50], Ye et al. present a contention-look-ahead on-chip routing scheme that is similar to [36]. It is a nonminimal routing scheme in the sense that, based on the value of two delay penalty indices, the router chooses whether to send the packet towards a *profitable route* (minimal route) or a *misroute* (nonminimal route). In [2], Ascia et al. presented a new selection strategy named *Neighbors-on-Path*, developed with the aim to choose output channels such that a packet can be routed to its destination along a path that is as free as possible of congested nodes.

5.2.4 Routing Logic and Hardware Implications

Depending on how the routing algorithm is implemented, its contribution to the overall power dissipation could be relevant. As far as routing function is concerned, it can be implemented in two main ways, namely *algorithmic* and *table based*. In algorithmic implementation, the routing function is described by means of *if-then-else* statements that are synthesized into a combinatorial logic circuit. Table-based implementation uses a routing table which stores routing information in the form of a set of entries (an entry for each destination), each containing the information about the output port (or the set of output ports) through which the packet must be forwarded to reach its destination.

   Each of them has its own pros and cons. An algorithmic implementation of the routing function is generally cheap in terms of both silicon area and power dissipation. But, it is not very flexible as it is tailored for a particular network topology. In addition, designing an algorithmic implementation of the routing function could be very difficult (sometime impossible) for nonhomegeneous topologies. On the other side, table-based implementation of the routing function is the most flexible solution since it can be updated at run-time (re-configuration) to be adapted to the particular scenario [15, 31], such as the changes in traffic conditions (e.g., multi use-case applications) and the variation of the network topology (e.g., due to transient or permanent faults), etc. But, as it will be shown in Sect. 5.6, silicon area occupancy and power dissipated by the routing table cannot be neglected. In addition, differently from the rest of the elements which form a router, its contribution becomes more and more important as the network size increases. However, it should be pointed out that the routing table is accessed, and thus dissipates power, only when the header flit is processed.

   If source routing is used, routing computation is performed only once, at the network interface, making routing logic inside the router extremely simple. Although the power dissipation of the routing logic in the router is minimized, the increase in packet size (as routing path information must be stored into the packet) could have negative impact on performance, resulting in an increase in the energy consumption.

5.2.5 Region Concept in NoC

The two-dimensional mesh topology is favored by several researchers, because of its good NoC design properties. Its natural match with 2D chip layout enables not only low cost but also high performing network structures. A main drawback with regularity is the inefficiency of handling cores of different sizes. Since a 2D mesh NoC implies links of equal length and core slots of equal size, it follows that the largest core determines the slot size. If the difference in core size is large, a lot of area will be wasted. The concept of region was proposed as a solution to this problem [28].

Figure 5.3 illustrates a nonhomogenous 2D mesh NoC with standard sized tiles and one larger region tile. Since the region covers several standard resource slots, it is assumed that routers cannot be placed within the region boundaries. The wrapper provides the interface between the communication protocols of the region core and the network. The access points are the network routers that directly connect the region via the wrapper. A number of applications of the region concept are discussed in [21]. A region may, for example, not only be considered for physically different structures. In this case, configurable routers dynamically create and maintain regions in a NoC, depending on application requirements.

Another application of the region concept is the encapsulation of areas in the NoC with specific communication requirements, such as power consumption or QoS. While there are several advantages of regions, quite a few drawbacks also exist [21]. One such drawback is the amplified difficulties of designing efficient deadlock-free routing algorithms.

Algorithms for irregular networks like Up*/Down* [43] are applicable, but their general properties often result in relatively lower efficiency on regular topologies.

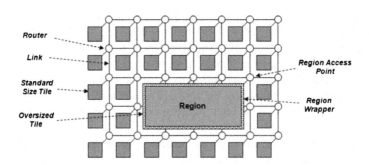

Fig. 5.3 NoC with Region

Several fault-tolerant routing algorithms for 2D mesh networks use a fault model (faulty blocks) [23, 49] that resembles a region. Typically, these are based on some simple regular algorithm, which is enhanced to enable deadlock-free circumvention of faulty blocks. Virtual channels are a frequently used, but still expensive, feature in this area [5]. Still, some proposals that do not require virtual channels have been proposed [7, 23].

5.2.6   Network Energy and Routing Algorithms

The routing algorithm influences the energy requirements of a network in several ways. One important energy aspect is the requirements of chip resources. These generally increase with complexity of the routing algorithm or with the use of other hardware features, such as virtual channels and router tables.

Resource requirements affect the average power dissipation and consequently complex routing algorithms will have a negative impact on router energy consumption. On the other hand, this drawback may be reversed if packets are routed more efficiently. Therefore, it is possible that a high-cost routing algorithm is more energy efficient than a low-cost algorithm if it achieves a lower average message latency. Note that, increased latency, as shown in [44] caused by voltage downscaling at lower router activity, may not outweigh the power benefits of these techniques.

Different types of energy models have been used for calculation of energy requirements of routing algorithms. These models are commonly based on static and dynamic power estimates from synthesized gate-level models of the network components. Static power is mainly dependent on the required chip resources, whereas dynamic power is determined by the switching activity.

Dynamic power dissipation of a router can, as listed by Wolkotte et al. [48], depend on average load, concurrent flows, control overhead, and bit-flips in the data streams. These parameters are used for defining ten different scenarios of router activity: from idle, where the router is not handling any traffic, to a case where the router is fully occupied. Another router power estimation approach is taken by Lee and Bagherzadeh [29], who propose a high level energy model based on regression analysis.

For estimation of total network energy consumption, including traffic, Ye et al. [50] use the notion of bit-energy $E_{bit} = E_{Wbit} + E_{Sbit}$. The bit energy E_{bit} is composed of energy on interconnect wires E_{Wbit}, which is assumed to be proportional to length (load capacitance) and energy within the switch logic E_{Sbit}. Switch energy E_{Sbit} is estimated by application of random data streams. The bit-energy is used for determining flit-energy per hop, which in turn is used for estimation of the network energy.

Kim et al. [27] use a more elaborate model and show that the duration of time packets spend in a network is important for energy estimation. The model is used in a comparison between an adaptive routing algorithm and dimension ordered routing. Even though the adaptive algorithm is more resource demanding, it provides lower message latency which ultimately results in lower energy requirements. A similar conclusion is made in [30], for an adaptive router with 24% higher power dissipation than a deterministic counterpart.

The multiplicity of routes from adaptive routing algorithms also allows for higher controllability of link load. In this area, Hu and Marculescu [25] show that mapping, using adaptive algorithms (odd-even and west-first), can reduce power consumption as compared to using the deterministic XY routing.

5.2.7 Common Performance Metrics

Several metrics have been used for estimating, evaluating and comparing the performance of various routing algorithms [41]. These metrics include different versions of latency (spread, minimum, maximum, average and expected), various versions of throughput, jitter in latency, jitter in throughput, etc. Adaptivity is another measure used in this context and refers to the ability of the algorithm to offer alternative paths for a packet. A highly adaptive routing algorithm has a potential of providing high performance (low latency, low packet drop and high throughput), fault tolerance and uniform utilization of network resources. High adaptivity increases the chances that packets may avoid hot spots or faulty components and reduces the chances that packets are continuously blocked. Higher adaptivity facilitates network management and allows to more evenly distribute the traffic over the network, which generally leads to higher performance for the applications [39].

5.3 Terminology and Definitions

In an embedded system scenario, the communication traffic between different cores of a system-on-a-chip is usually well characterized. In particular, after the task mapping phase of the NoC design flow, we have a complete knowledge about the pairs of cores which communicate and other pairs which never communicate. This additional information can be exploited to design an application-specific routing algorithm which is highly adaptive and is also deadlock free. This information can be incorporated in Duato's theory for systematic design of deadlock-free routing algorithms for communication networks [13]. In this section, the concept of *application-specific routing* is introduced.

5.3.1 Basic Definitions

Given a directed graph $G(V, E)$, where V is the set of vertices and E is the set of edges, we indicate with $e_{ij} = (v_i, v_j)$ the directed arc from vertex v_i to vertex v_j.

Given an edge $e \in E$, we indicate with $src(e)$ and $dst(e)$, respectively, the source and the destination vertex of the edge [e.g., $src(e_{ij}) = v_i$ and $dst(e_{ij}) = v_j$].

Definition 5.1. A communication graph, $CG = G(T, C)$ is a directed graph, where each vertex t_i represents a task and each directed arc $c_{ij} = (t_i, t_j)$ represents the communication from t_i to t_j.

Definition 5.2. A topology graph, $TG = G(P, L)$ is a directed graph, where each vertex p_i represents a node of the network and each directed arc $l_{ij} = (p_i, p_j)$ represents a physical unidirectional channel (link) connecting node p_i to node p_j.

Definition 5.3. A mapping function, $M : T \rightarrow P$ maps a task $t \in T$ on a node $p \in P$.

Let $L_{in}(p)$ and $L_{out}(p)$ be the set of input channels and output channels for node p respectively. Mathematically:

$$L_{in}(p) = \{l \mid l \in L \land dst(l) = p\},$$
$$L_{out}(p) = \{l \mid l \in L \land src(l) = p\}.$$

Definition 5.4. A routing function for a node $p \in P$ is a function $R(p) : L_{in}(p) \times P \rightarrow \wp(L_{out}(p))$. $R(p)(l, q)$ gives the set of output channels of node p that can be used to send a message received from the input channel l and whose destination is $q \in P$. We assume that $R(p)(l, q) = \emptyset$ if q is not reachable from p.

The $\wp$ indicates a power set. We indicate with R the set of all routing functions: $R = \{R(p) : p \in P\}$.

5.3.2 Channel Dependency Graph and Deadlock Freedom

We briefly report two fundamental definitions and one theorem given by Duato [13, 14], which are used in the rest of the chapter and extended for application specific context.

Definition 5.5. Given a topology graph, $TG(P, L)$ and a routing function R, there is a *direct dependency* from $l_i \in L$ to $l_j \in L$ if l_j can be used immediately after l_i by messages destined to some node $p \in P$.

Definition 5.6. A channel dependency graph $(CDG)(L, D)$ for a topology graph TG, and a routing function R, is a directed graph. The vertices of CDG are the channels of TG. The arcs of CDG are the pair of channels (l_i, l_j) such that there is a direct dependency from l_i to l_j.

The following theorem due to Duato is a straightforward extension of Dally and Seitz theorem [12] for adaptive routing functions.

Theorem 5.1 (Duato's Theorem [13]). *A routing function R for a topology graph TG is deadlock-free if there are no cycles in its channel dependency graph CDG.*

5.3.3 Application-Specific Channel Dependency Graph

The above definitions and theorem do not make any reference to the communication traffic because they implicitly assume that all the nodes of the network can communicate with each other. As stated at the beginning of the section, in an embedded system scenario, the designer often knows which nodes of the network communicate and which do not. This information can be captured by a communication graph CG. In this section, we extend Duato's theory by incorporating this communication topology information.

Definition 5.7. Given a communication graph $CG(T, C)$, a topology graph $TG(P, L)$, a mapping function M, and a routing function R, there is an *application-specific direct dependency* from $l_i \in L$ to $l_j \in L$ iff

$$dst(l_i) = src(l_j), \tag{5.1}$$

$$\exists c \in C : l_j \in R(dst(l_i))(l_i, M(dst(c))). \tag{5.2}$$

Condition (5.1) states that there exists a possibility for a message to use l_j immediately after l_i. Condition (5.2) states that there exists a communication that can actually use l_j immediately after l_i.

Definition 5.8. An application specific channel dependency graph (ASCDG)(L, D) for a given CG, a topology graph TG, and a routing function R, is a directed graph. The vertices of ASCDG are the channels of TG. The arcs of ASCDG are the pair of channels (l_i, l_j) such that there is an application-specific direct dependency from l_i to l_j.

In this chapter, we assume minimal routing while constructing ASCDG.

Theorem 5.2. *A routing function R for a topology graph TG and for a communication graph CG is deadlock-free if there are no cycles in its application-specific channel dependency graph ASCDG.*

Proof. As the ASCDG is acyclic, it is possible to establish an order between the channels of L. Suppose that there is a deadlocked configuration for R. Let l_i be a channel of L with a nonempty queue such that there are no channels less than l_i with a nonempty queue. The following two cases are possible:

Case 1: l_i is minimal. In this case, as proved in [13], l_i is a sink, i.e., a channel such that all the flits that enter on it reach the destination in a single hop. Then, the flit at the queue head can reach its destination in a single hop and there is no deadlock.

Case 2: l_i is not a minimal. For each $l_j \in L$ such that $l_i > l_j$, there are no flits in the queue for channel l_j. Thus, the flit at the head of the queue for l_i is not blocked, regardless of whether it is a header or a data flit, and there is no deadlock. $\square$

5.3.4 Routing Adaptivity

We use the definition of adaptivity given by Glass and Ni in [18], which is also used by Chiu in [10] and by many other researchers. We define adaptivity, $\alpha(c)$ (sometimes also referred as degree of adaptiveness), for a communication c as the ratio of the number of allowed minimal paths to the total number of minimal paths between the source node and the destination node:

$$\alpha(c) = \frac{|\Phi(c)|}{TMP(c)},$$

where $TMP(c)$ represents the total number of minimal paths from node $M(src(c))$ to node $M(dst(c))$. The average adaptivity, α, is the average of the degree of adaptiveness for all the communications:

$$\alpha = \frac{1}{|C|} \sum_{c \in C} \alpha(c).$$

5.4 APSRA Design Methodology

The APSRA design methodology is depicted in Fig. 5.4. It gets as inputs the application modeled as a task graph (or as set of concurrent task graphs) and the network topology modeled as a topology graph. We assume that the tasks in the application have already been mapped and scheduled on the available NoC resources. Using this information, APSRA generates a set of routing tables (one for each router of the NoC), which not only guarantee both reachability and deadlock freeness of communication among tasks but also try to maximize routing adaptivity. A compression technique can be used to compress the generated routing tables. Finally, the compressed routing tables are uploaded to the physical chip (NoC configuration). Of course, all the aforementioned steps are carried out offline.

Let us start with a brief overview of APSRA methodology by means of an example.

5.4.1 APSRA by Example

For the sake of example, let us consider the communication graph and the topology graph depicted in Fig. 5.5. Although for this example the topology is mesh-based, the approach is general and can be applied to any network topology without any modification. For the sake of simplicity, as mapping function let us consider $M(T_i) = P_i, i = 1, 2, \ldots, 6$.

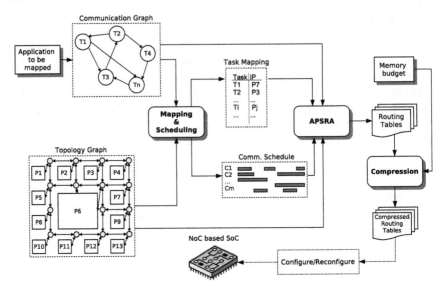

Fig. 5.4 Overview of the APSRA design methodology

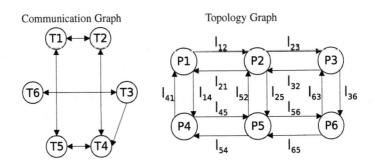

Fig. 5.5 Communication graph and topology graph

The CDG [14] for a minimal fully adaptive routing algorithm is shown in Fig. 5.6. Since it contains several cycles, Duato's theorem [14] cannot assure deadlock freedom for minimal fully adaptive routing for this topology. To make the routing deadlock free, it is necessary to break all cycles of the CDG. Breaking a cycle by means of a dependency removal results in restricting the routing function and, consequently, loss of adaptivity. As the cycles to be removed are many, the adaptivity of the resulting deadlock-free routing algorithm will be strongly reduced.

However, if communication information is considered, several cycles of the CDG can be safely removed without any impact on the adaptivity of the routing algorithm. For example, study the dependency $l_{1,2} \to l_{2,3}$. Such a dependency is present in the CDG but it is not present in the ASCDG. In fact, channels $l_{1,2}$ and $l_{2,3}$ can be used

Fig. 5.6 Channel dependency graph for the network topology in Fig. 5.5, given a minimal fully adaptive routing algorithm

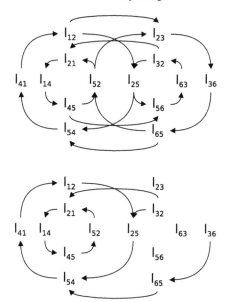

Fig. 5.7 Application-specific channel dependency graph for the network topology and the communication graph in Fig. 5.5, given a minimal fully adaptive routing algorithm

in sequence only for the communications $T_1 \rightarrow T_3$, $T_1 \rightarrow T_6$, and $T_4 \rightarrow T_3$ which are not present in the CG. If we analyze the rest of the dependencies taking into consideration the CG, we find that several dependencies can be safely removed without the need of restricting the routing function. The resulting $ASCDG$ is shown in Fig. 5.7. It has been obtained from the CDG by removing all the dependencies which cannot be activated by the communication graph.

Although the number of cycles is reduced to two for the $ASCDG$, we still have a possibility of deadlock. To handle this, we can simply break the cycles as follows. The application-specific channel dependency $l_{4,1} \rightarrow l_{1,2}$ is due to the communication $T4 \rightarrow T2$. Such communication can be realized by both paths $P4 \rightarrow P5 \rightarrow P2$ and $P4 \rightarrow P1 \rightarrow P2$. If the routing function is restricted in such a way that the latter path is prohibited, the application specific channel dependency $l_{4,1} \rightarrow l_{1,2}$ does not exist any longer. In a similar way, it is possible to break the second cycle, removing, for instance, the dependency $l_{1,4} \rightarrow l_{4,5}$ due to communication $T1 \rightarrow T5$.

However, this restriction reduces the degree of adaptiveness of the routing algorithm. Now suppose that we have some knowledge about communication concurrency and suppose that communication $T1 \rightarrow T5$ and communication $T2 \rightarrow T4$ do not overlap in time. Figure 5.8 highlights the dependencies due to such communications. Since these communications are not concurrent, the associated dependencies are not concurrently active too. The result is that the two cycles are actually false cycles. In conclusion, for this latter case, a minimal fully adaptive routing algorithm is deadlock free.

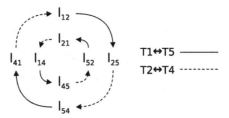

Fig. 5.8 The cycles in the application-specific channel dependency graph of Fig. 5.7 may be removed if the communications T1↔T5 and T2↔T4 are not concurrently active

```
 1  APSRA(in: CG, TG, M; out: RT)
 2  {
 3      R ← MinimalFullyAdaptiveRouting(CG,TG,M);
 4      BuildASCDG(CG,TG,M,R,ASCDG);
 5      GetCycles(ASCDG,C);
 6      RemoveCycles(C,ASCDG,CG,TG,M,R,success);
 7      if (success)
 8          ExtractRoutingTables(R,RT);
 9      else
10          RT ← ∅;
11  }
```

Fig. 5.9 APSRA main algorithm

5.4.2 Main Algorithm

The main algorithm which implements the APSRA methodology is shown in Fig. 5.9. It gets as inputs a communication graph, CG, a topology graph, TG and a mapping function M and returns the set of routing tables. The algorithm starts by initializing R with a minimal fully adaptive routing function then calls the procedure BuildASCDG which builds the *ASCDG*. The procedure GetCycles returns the set C of cycles in the ASCDG. RemoveCycles tries to remove all cycles in C from the ASCDG with the objective of minimizing the loss of adaptivity and with the constraint to guarantee reachability between all communicating pairs (cf. Sect. 5.4.3). If it succeeds, it returns *true* and the procedure ExtractRoutingTables is used to extract routing tables from R (cf. Sect. 5.4.4), otherwise an empty set is returned.

5.4.3 Cutting Edge with Minimum Loss of Adaptivity

As discussed above, the procedure RemoveCycles tries to remove all the cycles from the ASCDG with the objective of minimizing loss of adaptivity, subject to the constraint of guaranteeing reachability. Before we start presenting the heuristic used to select the application-specific channel dependency to be removed to break a cycle, some definitions should be given.

Given a communication $c \in C$, we denote with $\Phi(c)$ the set of all permissible minimal paths from node $M(src(c))$ to node $M(dst(c))$. We indicate with $\phi_i(c)$ the ith path of $\Phi(c)$. A *Path* from a node p_s to a node p_d is a succession of channels $\{l_1, l_2, \ldots, l_n\}$, $l_i \in L$ such that $dst(l_i) = src(l_{i+1})$, $i = 1, 2, \ldots, n - 1$, $p_s = src(l_1)$, and $p_d = dst(l_n)$.

Definition 5.9. For an edge d of the ASCDG the *Pass-through set* $A(d)$ is the set of pairs (c, j) where $c \in C$ is a communication whose jth path contains both channels associated to d. Formally:

$$A(d) = \{(c, j) | c \in C, \ j \ \in \mathbb{N} \ : \ src(d) \in \phi_j(c) \wedge dst(d) \in \phi_j(c)\}.$$

For an edge d of the ASCDG, we indicate as $A(d)|_c$ the restriction of $A(d)$ for communication $c \in C$, that is

$$A(d)|_c = \{(c', \cdot) \in A(d) : c' = c\}.$$

Theorem 5.3. *Given an ASCDG(L, D) and $d = (l_i, l_j) \in D$ then $A(d) \neq \emptyset$.*

Proof. $\Rightarrow d = (l_i, l_j) \in D$ then $\exists c \in C$ such that $l_j \in R(dst(l_i))(l_i, c)$ that is, there exists a communication c which has a path that contains both l_i and l_j. This path belongs to $\Phi(c)$ and suppose it is the jth path of $\Phi(c)$ named $\phi_j(c)$. Then the pair (c, j) belongs to $A(d)$ because $src(d) = l_i \in \phi_j(c)$ and $dst(d) = l_j \in \phi_j(c)$.

$\Leftarrow$ Let $a = (c, j) \in A(d)$, then $\exists \phi_j(c) \in \Phi(c)$ that contains both $src(d) = l_i$ and $dst(d) = l_j$. The condition (5.1) is satisfied by construction because $d \in D$. The existence of the path $\phi_j(c)$ states that a communication c travelling on l_i can immediately use l_j. This means that the routing function at node $dst(l_i)$ allows this turn. Therefore, $l_i \in R(dst(l_i))(l_i, M(dst(c)))$ and the condition (5.2) are satisfied too. □

Hence, let $D_c = \{d_1, d_2, \ldots, d_n\} \subseteq D$ be a cycle in the ASCDG(L, D). To break the cycle, a dependency d_i must be removed. If d_i is removed, based on Theorem 5.3, $A(d_i) = \emptyset$. To make $A(d_i) = \emptyset$, the set of admissible paths for some communications has to be restricted. This, however, has an impact on the degree of adaptiveness of the routing function. The heuristic has to select the dependency d_i to be removed in such a way that the impact on the degree of adaptiveness is minimized.

Let α be the current degree of adaptiveness and α_d the degree of adaptiveness when we remove a dependency $d \in D_c$. The objective is to minimize the difference $\alpha - \alpha_d$, or equivalently maximize α_d.

$$\max_{d \in D_c} \alpha_d = \max_{d \in D_c} \frac{1}{|C|} \sum_{c \in C} \frac{|\Phi_d(c)|}{TMP(c)}, \tag{5.3}$$

where with $\Phi_d(c)$, we indicated the set of paths for communication c when the dependency d is removed, that is

$$\Phi_d(c) = \Phi(c) \setminus \{\phi_j(c)|\ (c, j) \in A(d)_{|c}\},$$

we have:

$$|\Phi_d(c)| = |\Phi(c)| - |\{\phi_j(c)|\ (c, j) \in A(d)_{|c}\}|$$
$$= |\Phi(c)| - |A(d)_{|c}|.$$

Substituting into (5.3), we have:

$$\max_{d \in D_c} \frac{1}{|C|} \left(\sum_{c \in C} \frac{|\Phi(c)|}{TMP(c)} - \sum_{c \in C} \frac{|A(d)_{|c}|}{TMP(c)} \right),$$

which is equivalent to:

$$\min_{d \in D_c} \sum_{c \in C} \frac{|A(d)_{|c}|}{TMP(c)} = \min_{d \in D_c} \sum_{(c, \cdot) \in A(d)} \frac{1}{TMP(c)}.$$

In short, the heuristic states that to minimize the impact on adaptiveness we have to select a dependency $d \in D_c$ to be removed which satisfy the following reachability constraint:

$$\bigwedge_{c \in C(d)} |\Phi(c)| > |A(d)_{|c}|, \tag{5.4}$$

and minimize the quantity:

$$\sum_{(c, \cdot) \in A(d)} \frac{1}{TMP(c)}. \tag{5.5}$$

Where with $C(d)$, we indicate the set of communications having at least a path which contains both channels associated with d. The inequality (5.4) ensures that all the communications which use the links $src(d)$ followed by $dst(d)$ will have alternative paths after d is removed. The removal of a dependency d impacts on $\Phi(c)$ as follows:

$$\forall (c, j) \in A(d) \Rightarrow \Phi(c) = \Phi(c) \setminus \{\phi_j(c)\}.$$

Restricting the routing functions in various network nodes may also affect reachability of certain communications. The order in which the cycles in *ASCDG* get treated may finally decide whether the constraint (5.4) can be met for all cycles or not. This implies that if we look at cycles in one order only, then we may not get a routing path for some communications. In fact, in the worst case, we may have to exhaustively consider all possible combinations of dependencies, one from each cycle in *ASCDG*, to be removed to find a feasible minimal routing for all communicating pairs.

5.4.4 Routing Tables

For each node $p \in P$, and for each input channel $l \in L_{in}(p)$, there is a routing table $RT(p,l)$ in which each entry consists of (1) a *destination address $d \in P$* and (2) a set of output channels $O \in \wp(L_{out}(p))$ that can be used to forward a message received from channel l and destined to node d. Formally

$$RT(p,l) = \{(d,O)|\, d \in P,\ O = R(p)(l,d) \wedge O \neq \emptyset\}.$$

The routing table of a node $p \in P$ is the union of routing tables of each input channel of p:

$$RT(p) = \bigcup_{l \in L_{in}(p)} RT(p,l).$$

Although in general table-based implementation of routing function is more costly in terms of silicon area as compared to a custom logic implementation, it can be a blessing. In fact, table-based implementation allows configurability (and even dynamic re-configurability [15, 31]) of the routing algorithm to allow modifications in communication requirements in the running applications. However, routing table size can be drastically reduced as we will see in Sect. 5.4.4.1.

5.4.4.1 Routing Table Compression

Although routing tables represent the most flexible way to implement a routing function, as discussed in Sect. 5.2.4, they have strong implications on router's cost (i.e., silicon area) and power dissipation. In order to attenuate this impact, a method for router table compression was proposed in [37]. The method, developed for 2D mesh topology, allows for reducing the size of routing tables such that the resulting routing function remains deadlock free as well as highly adaptive.

The basic idea of the compression method is shown in Fig. 5.10. Let us suppose that the routing table associated with the west input port of node X is that shown in Fig. 5.10a. Since we are considering minimal routing, the destination node for a packet received from the west port will be in the right part of the NoC.

Now, let us associate a color with each one of the admissible output combinations in the routing table of Fig. 5.10a. For instance, as shown in Fig. 5.10b, we mark with colours Green and Blue all the destination nodes which need that a packet at the west input port must be forwarded through the South and East output port respectively.

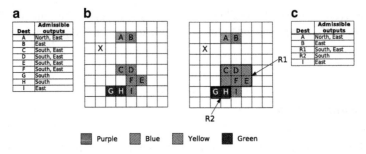

Fig. 5.10 The basic idea of the routing compression method. (**a**) Routing table before compression. (**b**) Colour based clustering. (**c**) Compressed routing table

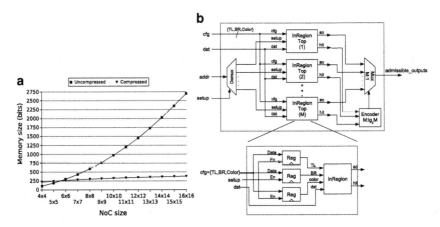

Fig. 5.11 Router table sizes (in bits) for the uncompressed and compressed cases for different mesh sizes (**a**) Block diagram of the architecture implementing the routing function which uses the compressed routing table (**b**)

Note that the rest of the destination nodes can be reached from multiple outputs. Nodes which can be reached from North or East (South or East) output ports are marked with Purple (Yellow) colour.

The next step is to perform a colour-based clustering of the destination nodes by means of rectangular regions. Right side of Fig. 5.10b shows an example of such a clustering. The basic idea of the compression technique is that it is no more necessary to store the set of all destinations but only the set of clusters as shown in Fig. 5.10c.

Still, a compressed-table router requires some extra bits for holding attributes characterizing the cluster regions (i.e., identification by the top left and bottom right nodes and colour). Figure 5.11a shows, for different mesh sizes, the total number of bits to store in a router for both the uncompressed and the compressed routing table with four regions per input.

If a small loss of adaptivity is allowed, a cluster merging procedure can be used to increase the compression ratio. For instance, from Fig. 5.10c, clusters R1 (yellow) and R2 (green) can be merged to form a single cluster R3 (yellow) whose destinations can be reached using only the east output port. As reported in [37], in practical cases, using four to eight clusters is enough to cover all practical traffic scenarios without any loss of adaptivity. Using two to three clusters gives a small reduction in adaptivity which translates into less than 3% performance degradation as compared to using uncompressed routing tables.

From the architectural viewpoint, the logic to manage the compressed routing table is very simple. Figure 5.11b shows a block diagram of the architecture implementing the routing function which uses the compressed routing table. For a given destination *dst*, the block extracts from the compressed routing table the set of admissible outputs. The input port *setup* allows to configure the router. When *setup* is asserted *addr* input is used to select the region to be configured. Configuring a region means storing the region attributes (top left, bottom right and colour) into the compressed routing table.

The block *InRegion* checks if a destination *dst* belongs to a region identified by its top left (TL) corner and its bottom right (BR) corner. If this condition is satisfied, the output *ao* assumes the value of the colour input and the output hit is set. The advantages of using compression techniques in terms of both silicon area and power dissipation will be quantitatively discussed in Sect. 5.6.1.

5.5 Performance Evaluation of APSRA

This section analyzes the performance of APSRA in terms of both adaptivity and simulation-based evaluation of latency and throughput for both homogeneous and nonhomogeneous mesh NoC with regions. The experimental flow is as follows. The communication graph is extracted from the traffic scenario that characterizes the application. The communication graph and the topology graph (i.e., the network topology) are the inputs of the APSRA procedure reported in Fig. 5.9. The APSRA builds the ASCDG and, if it contains cycles, the heuristic presented in Sect. 5.4.3 is iteratively used until the ASCDG becomes acyclic. Then, routing tables are generated and the degree of adaptiveness of the routing function is calculated. Finally, routing tables are used by the NoC simulator [16] to analyze the dynamic behavior of the generated routing algorithm.

5.5.1 Traffic Scenarios

Performance evaluation is carried out on both synthetic and real traffic scenarios. A brief description of them is as follows.

- *Random* (a.k.a. Uniform): For a given source node, the destination node is selected randomly. The number of vertices (tasks) of the communication graph is

fixed, whereas the number of edges (communications) is a user-defined parameter. We define the *communication density*, ρ, as the ratio between the number
of communications and the number of tasks. Thus, the communication graph
associated with the random traffic with a communication density of ρ is a randomly generated graph with the number of vertices equal to the number of nodes
of the network. Each vertex has on average ρ outgoing edges directed to other,
randomly selected, vertices of the graph.

- *Locality*: It is similar to *Random* traffic with the difference that, in this case,
  the probability of selecting a destination node is related to its distance from the
  source node. More precisely, we define the *one-hop probability*, ohp, as the communication probability between two tasks mapped on two nodes distant one hop
  from each other. The communication probability, CP, of two tasks mapped on
  two nodes distant $h \geq 2$ hops is computed as $CP(h) = (1 - \sum_{i=1}^{h-1} CP(i))/2$,
  where $CP(1) = ohp$.
- *Transpose 1*: A node (i, j) only sends packets to a node $(N - 1 - j, N - 1 - i)$,
  where N is the size of the mesh.
- *Transpose 2*: A node (i, j) only sends packets to a node (j, i).
- *Hotspot-4c*: Some nodes are designated as *hot spot nodes*, which receive hot
  spot traffic in addition to regular uniform traffic [4]. Given a hot spot percentage h, a newly generated packet is directed to each hot spot node with
  an additional h percent probability. In *Hotspot-4c*, the hot spot nodes are located at the centre of the mesh $[(3, 3), (4, 3), (3, 4), (4, 4)]$ with 20% hot spot
  traffic.
- *Hotspot-4tr*: Hot spot traffic in which the hot spot nodes are located at the
  top-right corner of the mesh $[(0, 6), (0, 7), (1, 6), (1, 7)]$ with 20% hot spot
  traffic.

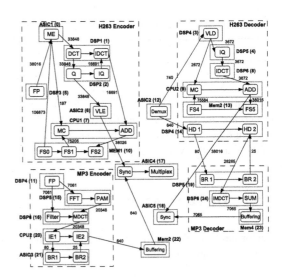

Fig. 5.12 Communication
graph of a generic multimedia
system [25]

- *Hotspot-8r*: Hot spot traffic in which the hot spot nodes are located at the right side of the mesh $[(i, 7), i = 0, 1, \ldots, 7]$ with 10% hot spot traffic.
- *MMS*: Traffic generated by a generic MultiMedia System which includes an H.263 video encoder, an H.263 video decoder, an mp3 audio encoder, and an mp3 audio decoder [25] where the communication graph is shown in Fig. 5.12. The application is partitioned into 40 distinct tasks, which then were assigned and scheduled onto 25 selected IPs. The topological mapping of IPs into tiles of a 5×5 mesh-based NoC architecture has been obtained by the approach presented in [1].

5.5.2 Adaptivity Analysis

This subsection compares APSRA with adaptive routing algorithms based on the *turn model* [18] and with the *Odd-Even* turn model [10] in terms of adaptivity. Figure 5.13a shows the average degree of adaptiveness for different NoC size and for $\rho = 2$ (top) and $\rho = 4$ (bottom). Each point of the graph has been obtained by evaluating 100 random communication graphs and reporting the mean value and the 90% confidence interval. The algorithms based on turn model outperform in degree of adaptiveness for large mesh size. Unfortunately, the degree of adaptiveness provided by the turn model is highly uneven [10]. This is because at least half of the source-destination pairs are restricted to having only one minimal path, while full adaptiveness is provided for the rest of the pairs. This is confirmed by the high standard deviation values these algorithms exhibit (Fig. 5.13b). On the other hand, Odd-Even is the worst one in terms of average degree of adaptiveness, but it is more even for different source–destination pairs. APSRA outperforms the other algorithms for small NoC size, but performance decreases very fast as NoC size increases and communication density increases. In fact, for a given network size, increasing communication density is the same as moving from the application-specific domain to the general purpose domain, where, in the extreme case, each node communicates with each other node of the network, and where the turn model gives an upper bound on maximum reachable degree of adaptiveness [18].

At any rate, the uniform random traffic scenario is not very representative for a NoC system [4]. Usually, in fact, cores that communicate most are mapped close to each other [1, 25, 34]. Figure 5.14 shows results obtained for locality traffic with $ohp = 0.4$. In this case, APSRA outperforms the other algorithms both in terms of adaptiveness (Fig. 5.14a) and standard deviation (Fig. 5.14b). Quantitatively, APSRA provides very high level of adaptivity, for $\rho = 2$ on average over 10% and 18% compared to turn model-based algorithms and Odd-Even, respectively. And for $\rho = 4$, it provides 7 and 15% higher adaptivity over turn model-based algorithms and Odd-Even, respectively. Moreover, the degree of routing adaptiveness provided by APSRA is more even for different source–destination pairs.

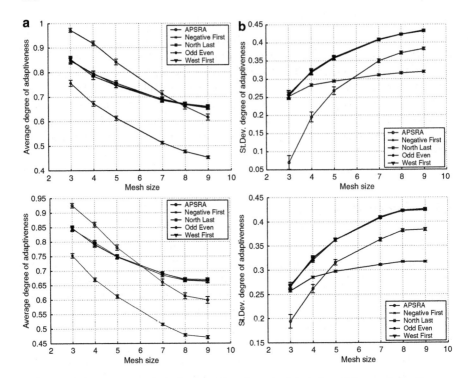

Fig. 5.13 Average degree of adaptiveness (**a**) and standard deviation of degree of adaptiveness (**b**) for different NoC sizes and for random generated communication graph with $\rho = 2$ (*top*) and $\rho = 4$ (*bottom*)

5.5.3 Simulation Based Evaluation

This subsection compares APSRA with both a deterministic routing algorithm (XY) and an adaptive routing algorithm (Odd-Even) using *throughput* and *delay* as performance metrics. Odd-Even is chosen as it has been proven to exhibit the best average performance with different traffic scenarios [10]. It is also a highly cited adaptive routing algorithm for mesh networks without virtual channels.

The evaluations are carried out on a 8×8 network using wormhole switching with a packet size randomly distributed between 2 and 16 flits, and routers with input buffer size of 2 flits. The maximum bandwidth of each link is set to 1 flit per cycle. We use the source packet injection rate (*pir*) as load parameter with Poisson packet injection distribution. In the case of *MMS* traffic, we instead consider self-similar packet injection distribution. Self-similar traffic has been observed in the bursty traffic between on-chip modules in typical MPEG-2 video applications [47] and networking applications [3]. For each load value, latency values are averaged over 60,000 packet arrivals after a warm-up session of 30,000 arrived packets. The 95%

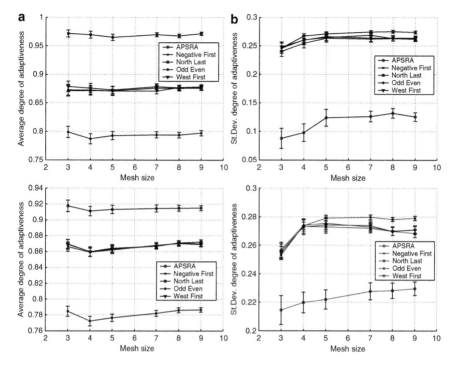

Fig. 5.14 Average degree of adaptiveness (a) and standard deviation of degree of adaptiveness (b) for different NoC size and for random generated communication graph with $\rho = 2$ (top) and $\rho = 4$ (bottom) and $ohp = 0.4$

confidence intervals are mostly within 2% of the means. For adaptive routing algorithms, two different selection policies are considered: *random* and *buffer level*. If multiple output ports are available for a header flit, a random selection policy selects the output randomly, whereas the buffer level policy selects the output whose connected input port has the minimum buffer occupied.

5.5.3.1 Homogeneous 2D Mesh NoC

Figure 5.15 (top) shows the results obtained under uniform traffic. From Fig. 5.15a we observe that the nonadaptivity of the XY algorithm results in a higher saturation point as compared to other algorithms. The main reason for this is that the XY algorithm inherently captures long-term global information [18]. By routing packets along one dimension first and then the other, the algorithm distributes traffic in the most uniform manner possible in the long term. Adaptive algorithms, on the other hand, select the routing paths on the basis of short-term local information. Their way of operating tends to create "zigzag" paths which hinder the uniform distribution of traffic, causing a greater channel contention that deteriorates performance at higher *pir* rates.

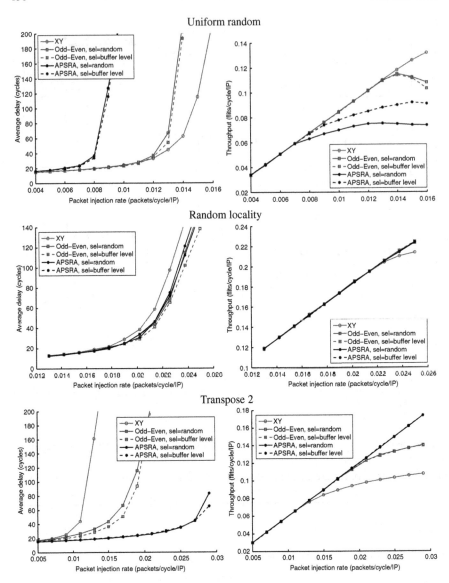

Fig. 5.15 Delay variation (*left*) and throughput variation (*right*) under uniform random traffic, random locality traffic with $\rho = 2$ and *ohp*= 0.4, and *Transpose 2* traffic

Figure 5.15 (middle) shows average delay and throughput for locality traffic with *ohp*= 0.4 and $\rho = 2$. The high adaptiveness exhibited by APSRA for this traffic scenario (see Fig. 5.14) does not translate into an improvement in delay and throughput. This is due to the fact that the communicating pairs are very close to each other and the number of alternative paths is not so high compared to the single path required by a deterministic routing algorithm.

For *Transpose 1* and *Transpose 2* traffic scenarios, APSRA outperforms the other routing strategies. Figure 5.15 (bottom) shows the results for *Transpose 2* (similar results have been obtained for *Transpose 1*). Maximum load that the network is capable of handling using a deterministic routing is 0.012 packets/cycle/IP. This value increases to 0.017 packets/cycle/IP when Odd-Even is used and reaches 0.028 packets/cycle/IP with APSRA.

Figure 5.16 shows average communication delay and throughput variation under hot spot traffic. In all three scenarios, APSRA outperforms the other algorithms. In particular, the difference between deterministic and adaptive routing algorithms is more evident for the second and third scenario. In fact, an 8×8 NoC with a hot spot node located at the center can be seen as a set of four small and isolated 4×4 NoCs (one NoC for each quadrant), each one stimulated with hot spot traffic where the hot spot node is located at one of the four corners. Since adaptiveness is really exploited in a large NoC, it justifies this behavior.

Finally, Fig. 5.17 shows the average delay and throughput variation for different injection loads under MMS traffic. For an injection load of 0.018 packets/cycle/IP, which is below saturation for all routing algorithms, the average delay is 36 cycles for XY, 32 cycles for Odd-Even, and 26 cycles for APSRA when a random selection policy is used. Using a selection policy based on buffer levels, average delay decreases to 30 cycles for Odd-Even and to 23 cycles for APSRA.

Tables 5.1 and 5.2 report a summary of the results, both in terms of average delay and maximum injection rate sustainable by the network. As regards the average delay, it has been measured at a packet injection rate where none of the algorithms are saturated[1]. The maximum packet injection rate sustainable by the network is the minimum *pir* which saturate the network. Saturating injection rate has been calculated as the *pir* at which throughput drops more than 5% from the earlier averaged slopes.

For each traffic scenario and for each routing algorithm, Table 5.1 reports the packet injection rate at the saturation point. It also shows the percentage of improvement for APSRA over XY and Odd-Even. On average, APSRA outperforms deterministic XY routing by 55% and adaptive Odd-Even routing by 27%.

Table 5.2 reports for each traffic scenario and for each routing algorithm, the average delay measured at an injection load below saturation. It also shows the percentage of improvement for APSRA over XY and Odd-Even. On average, APSRA outperforms both deterministic XY and adaptive Odd-Even by almost 50 and 30%, respectively, in terms of average delay.

5.5.3.2 NonHomogeneous 2D Mesh NoC with Regions

This subsection analyzes the performance of APSRA for 2D mesh NoC with regions [28]. A region is an oversized resource slot which may be defined in 2D

---

[1] A network is said to start saturating when increase in applied load does not result in linear increase in throughput [41].

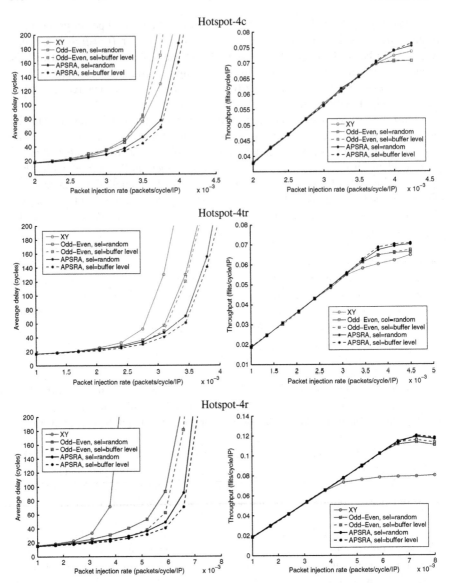

Fig. 5.16 Delay variation (*left*) and throughput variation (*right*) under *hot spot* traffic. Hot spot nodes at the center of the network (Hotspot-4c), at the *top right* corner of the mesh (Hotspot-4tr) and at the *right* side of the mesh (Hotspot-4r)

mesh topologies for support of, for example, larger cores or special communication requirements. However, forming a region in a regular 2D mesh will remove routers and links and the resulting topology will then be partially irregular (or nonhomogenous) [28]. For a short overview of regions in NoC, see Sect. 5.2.5.

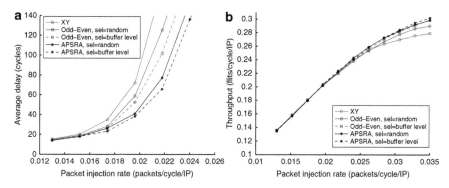

Fig. 5.17 Delay variation (**a**) and throughput variation (**b**) under traffic generated by a multimedia system

Table 5.1 Improvement in saturation point of APSRA as compared to XY and odd-even for different traffic scenarios

Traffic scenario	Max. *pir* (packets/cycle/IP)			APSRA improvement	
	XY	OE	APSRA	vs. XY(%)	vs. OE(%)
Random	0.0120	0.0105	0.0080	−33	−23
Locality	0.0190	0.0200	0.0210	10.5	5.0
Transpose 1	0.0110	0.0150	0.0270	145.5	80.0
Transpose 2	0.0110	0.0160	0.0270	145.5	68.8
Hotspot-4c	0.0033	0.0035	0.0038	13.6	7.1
Hotspot-4tr	0.0027	0.0031	0.0035	29.6	12.9
Hotspot-8r	0.0039	0.0059	0.0067	71.8	13.6
MMS	0.0174	0.0174	0.0196	12.6	12.6
Average improvement				54.6%	26.8%

Table 5.2 Improvement in average delay of APSRA as compared to XY and odd-even for different traffic scenarios

Traffic scenario	*pir* (pkts/cyc/IP)	Avg. delay (cycles)			APSRA impr.	
		XY	OE	APSRA	vs. XY(%)	vs. OE(%)
Random	0.007	18	18	23	−27	−27
Locality	0.020	39	34	29	24.2	13.2
Transpose 1	0.011	91	39	19	79.2	51.1
Transpose 2	0.011	82	31	19	76.6	38.4
Hotspot-4c	0.003	46	50	34	26.7	32.1
Hotspot-4tr	0.003	52	37	30	42.2	17.5
Hotspot-8rs	0.003	34	25	20	41.8	21.5
MMS	0.018	36	30	23	36.1	23.3
Average improvement					47.1%	28.7%

Since regular topology algorithms (e.g., X-Y, Odd-Even) cannot be used in this case, APSRA is here compared with a fixed version of Chen and Chiu's algorithm [22] which is adopted from the area of fault-tolerant routing. Experiments are carried out on a 7 × 7 mesh topology NoC with regions. The destinations for generated packets are randomly selected with hot spot probability of 60% for region access points. APSRA and Chen and Chiu's algorithm are evaluated using a region of size 2 × 2, either in the bottom left corner with three access points (bl_ap3) or in the centre of network with four access points (c_ap4).

Communication traffic is classified into three types, namely, as communication *traffic to region*, as *other traffic* where a resource other than the region is a destination, and as *all communications* which is the aggregate of the first two types of traffic. The first result shows average latency (delay) for all communications in the network, as depicted in Fig. 5.18a. The lowest latency values are obtained for APSRA with central region (apsra_c_ap4). Second lowest latency values are obtained with Chen and Chiu's algorithm and central region (chiu_c_ap4). After this comes APSRA with region in bottom left corner (apsra_bl_ap). The worst performance is shown by Chen and Chiu's algorithm and region in bottom-left corner (chiu_bl_ap3).

Figure 5.18b gives average latency for traffic with destinations other than the region. The worst latency, up to an injection rate of 5%, is obtained with Chen and Chiu's algorithm and region in centre (chiu_c_ap4). In this case, all the other combinations provide similar latency values in this range. However, when the injection rate is increased above 5%, the latency for Chen and Chiu's algorithm and region in corner position (chiu_bl_ap3) rapidly saturates. Next to saturate is APSRA with region in corner (apsra_bl_ap3). The best result from saturation point of view is shown for APSRA and the region in centre (apsra_c_ap4), although it has slightly higher latency at lower injection rates.

The results for traffic destined only to region is shown in Fig. 5.18c. Also in this case, APSRA with central region shows the best performance results in terms of low latency. Still, Chen and Chiu's algorithm with central region clearly gives better results than both algorithms achieve with the region at bottom-left position. Worst performance is also in this case shown by Chen and Chiu's algorithm with region in bottom-left corner.

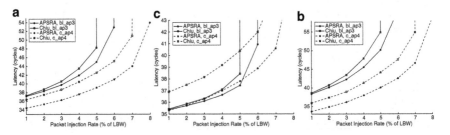

Fig. 5.18 Average latency vs. packet injection rate in % of link bandwidth with region in *bottom left* and *centre*. For all communications (**a**). For communications destined outside region (**b**). For communications destined to region (**c**)

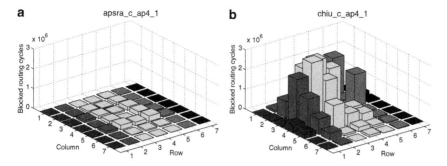

Fig. 5.19 Blocked routing cycles/router with (**a**) APSRA algorithm and (**b**) Chen and Chiu's algorithm

Figure 5.19 provides information that can explain the, sometimes large, differences in average latency. The diagrams present, for each network router, the total number of routing cycles when packets were blocked from advancing. These results are collected from simulations with $pir = 0.1$. Note that the scale of blocked routing cycles is not the same in the two diagrams.

When studying Fig. 5.19a, b, it is evident that the APSRA algorithm does not cause as much blockage as does Chen and Chiu's algorithm. Note that Chen and Chiu's algorithm results in more blockages close to the north and the west border of the region. The reason is that the paths in these areas are highly used by the algorithm in the procedures of routing around the region border. APSRA, on the other hand, is not biased towards specific routes and thus spreads the traffic more evenly around the border. As APSRA in many situations provides several paths to select from, packets often have the possibility to avoid congested routes which further decreases the blockage.

5.6 Cost, Power Dissipation and Energy Consumption Analysis

Till now, routing algorithms have been compared considering several metrics, such as average communication delay, throughput and adaptivity, which are mainly related to performance figure of merit of a communication system. This subsection compares routing algorithms considering two other important metrics, namely silicon area and power dissipation.

5.6.1 Generic Router Architecture

Before to start presenting silicon area and power figures, let us introduce the basic router architecture considered in this chapter. Figure 5.20 shows a block diagram

Fig. 5.20 Block diagram
of a generic router
architecture for mesh
topologies

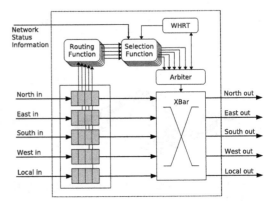

of a generic router architecture for mesh topologies. The router has six main inputs and five main outputs. North, east, south, west, and local in/out are the input/output ports used to connect the router with neighbor routers (north, east, south, and west directions) and with the local core. The input *network status information* is used by the selection function when an adaptive routing algorithm is considered. It conveys network status information, like, for instance, the buffer occupancy levels of the input buffers of the neighbor routers when buffer level selection policy is used. Of course, if a deterministic routing algorithm is considered, the selection function module is not present. The WHRT is the wormhole reservation table (and the associated logic) which is used to implement the wormhole switching technique. With exception of the selection function module, all the other elements shown in Fig. 5.20 are present in any router design for a deterministic routing algorithm or an adaptive routing algorithm. The router architectures considered in the rest of the section use a pipeline with the following three stages: FIFO, Routing/Selection, Arbitration/Crossbar.

5.6.2 Area and Power Dissipation

Figure 5.21 shows the area breakdown of seven wormhole-based routers implementing different routing algorithms. The values refer to a 90 nm technology. All the implementations use 64-bit flits and 4-entries FIFO buffers. In particular, XY implements the XY deterministic routing algorithm. Odd-Even (rnd) and Odd-Even (bl) implement Odd-Even routing algorithm with random and buffer-level selection policy, respectively. APSRA (rnd) and APSRA (bl) use a routing table with 64 entries (for a 8 × 8 mesh) and a random and buffer-level selection policy, respectively. Finally, APSRA-C uses compressed routing tables and the accompanying decoding logic to implement the routing function as described in [33]. As can be observed, the major contribution of silicon area is due to the FIFO buffers. In fact, they account

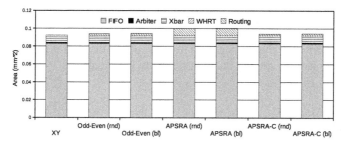

Fig. 5.21 Area breakdown of routers implementing different routing algorithms

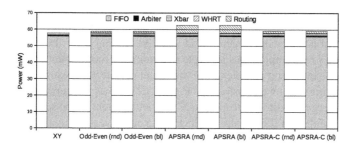

Fig. 5.22 Power breakdown of routers implementing different routing algorithms

for about 80–90% of the overall silicon area depending on the implementation. Crossbar and routing logic (routing function and selection policy) account for a non-negligible fraction of the router area. For table-based implementations, routing tables take almost 10% of the router area. However, this percentage can be significantly reduced using compression techniques. Precisely, applying the compression technique proposed in [37] to the routing tables generated by APSRA for the *Transpose 1* traffic scenario and constraining the compressed routing tables to 8 entries, it is possible to reduce the area of routing table to less than 2% of the overall router area. Overall, taking XY as baseline implementation, Odd-Even routers (both rnd and bl) are 3% more expensive. Routing table-based implementation of APSRA is about 8 and 9% more expensive than XY for random and buffer-level selection policy, respectively. If routing table compression [37] is used, the routing table overhead reduces to less than 3% [33].

With regard to power dissipation, Fig. 5.22 shows the averge power dissipation at 1 GHz for several router implementations. The power figures have been computed by simulating each element of the router separately using random stimuli. For this reason, the amount of power dissipated by a particular module (FIFOs, crossbar, arbitration, etc.) instantiated in different routers is the same for all routing algorithms. Once again the main power contribution is due to the FIFO buffers. Odd-Even routers are about 2% more power hungry than the XY router. Table-based implementations of routing algorithms increase power dissipation by 9%. However, using compression techniques, it is possible to combine the flexibility of table-based implementations with the economy (in terms of power dissipation) of an algorithic implementation of the routing function similar to Odd-Even.

5.6.3 Energy Consumption

It has been shown that, in general, adaptive routing algorithms outperform deterministic routing algorithms in both average communication delay and throughput under different traffic scenarios, both synthetic and captured from real applications. However, it has been observed that such improvements do not come for free. In fact, the architecture of a router implementing an adaptive routing algorithm is more complex than one which implements a simple deterministic routing function (Figs. 5.21 and 5.22). The increased complexity of the router architecture results in larger silicon area and higher power dissipation. Figure 5.23 shows the percentage decrease of average energy per flit with respect to XY under different traffic scenarios. As can be observed, although the routers implementing adaptive routing algorithms dissipate more power, the total message energy decreases because a given amount of traffic is drained in shorter time. On average, a general purpose adaptive routing algorithm like Odd-Even is about 14% more energy efficient than a deterministic routing algorithm. The improvement in energy consumption increases up to 36% when an application specific routing algorithm is considered.

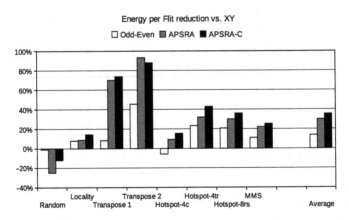

Fig. 5.23 Percentage decrease of average energy per flit with respect to XY under different traffic scenarios

5.7 Conclusions

NoC research is now entering a new phase where the emphasis will simultaneously be on low cost, high communication performance and low energy consumption. Reconfigurable platform chips, allowing static or dynamic configuration of resources and communication infrastructure based on the application requirements, is likely to lead the way forward. This chapter makes a contribution towards this direction. We have proposed a methodology, called APSRA, to develop efficient application-specific deadlock-free routing algorithms for networks on chip, which can easily be configured in routers. APSRA increases the attainable adaptivity by combining the channel dependency graph with a communication graph, which captures the communication among task in a NoC. The basic APSRA methodology is topology agnostic and is based on channel dependency graphs which are derived from network topology. The methodology described in the chapter uses only the information about topology of communicating cores. It has already been extended to exploit information about concurrency and bandwidth of communication transactions [38, 40]. Results based on the analysis and simulation-based evaluation demonstrate that routing algorithms developed using our approach significantly outperform general purpose routing algorithms such as XY and Odd-Even for mesh topology NoC, in terms of adaptivity and average latency. We also show that routing algorithms generated by APSRA methodology provide significantly higher performance over other deadlock-free routing algorithm for nonhomogeneous mesh NoCs. The main disadvantage of APSRA is that it requires table-based router design, which can be costly in terms of area as well as power dissipation. We have also proposed methods to compress routing tables thus reducing this disadvantage. Our analysis shows that at higher traffic loads the improvement in average latency translates into reduction in overall energy consumption also.

APSRA methodology uses a heuristic to remove minimum number of edges in the channel dependency graph to ensure deadlock freedom and minimum loss of adaptivity. The heuristic does not guarantee an optimal routing algorithm and, in some cases, may not be able to find a routing algorithm, that can route all messages. Therefore, there is a scope of improving this heuristic to get even higher adaptivity of the resultant routing algorithm and ensuring routing completeness. There is also a scope for developing new edge-removing heuristics, which have the goal of minimizing energy consumption or power dissipation rather than maximizing adaptivity. Developing multi-objective edge cutting heuristics which minimize energy consumption together with maximizing adaptivity will be an even more challenging problem. On-chip heat management is likely to become another important issue along with power dissipation for highly dense future chips. It will be important to avoid some chip subareas or some components on the chip becoming overheated. In the NoC context, routers and links are candidates for becoming hot spots. There is a scope for developing application-specific routing algorithms, which can avoid creating "temperature hot spots" on the chip.

Since APSRA methodology is application oriented and promises high performance and low energy consumption, we feel that it will be useful towards the development of configurable SoC platforms.

References

1. Ascia, G., Catania, V., Palesi, M.: Multi-objective mapping for mesh-based NoC architectures. In: Second IEEE/ACM/IFIP International Conference on Hardware/Software Codesign and System Synthesis, pp. 182–187. Stockholm, Sweden (2004)
2. Ascia, G., Catania, V., Palesi, M., Patti, D.: Implementation and analysis of a new selection strategy for adaptive routing in networks-on-chip. IEEE Transactions on Computers 57(6), 809–820 (2008)
3. Avresky, D.R., Shubranov, V., Horst, R., Mehra, P.: Performance evaluation of the ServerNetR SAN under self-similar traffic. In: International Symposium on Parallel and Distributed Processing, pp. 143–149 (1999)
4. Boppana, R.V., Chalasani, S.: A comparison of adaptive wormhole routing algorithms. In: International Symposium on Computer Architecture, pp. 351–360. San Diego, CA (1993)
5. Boppana, R.V., Chalasani, S.: Fault-tolerant wormhole routing algorithms for mesh networks. IEEE Transactions on Computers 44(7), 848–864 (1995)
6. Chatha, K., Srinivasan, K., Konjevod, G.: Approximation algorithms for design of application specific network-on-chip architectures. IEEE Transactions on Computer Aided Design of Integrated Circuits and Systems (2008)
7. Chen, K.H., Chiu, G.M.: Fault-tolerant routing algorithm for meshes without using virtual channels. Journal of Information Science and Engineering 14(4), 765–783 (1998)
8. Cherkasova, L., Kotov, V., Rokicki, T.: Fibre channel fabrics: Evaluation and design. In: Hawaii International Conference on System Sciences, pp. 53–58 (1996)
9. Chien, A.A., Kim, J.H.: Planar-adaptive routing: Low-cost adaptive networks for multiprocessors. Journal of the ACM 42(1), 91–123 (1995)
10. Chiu, G.M.: The odd-even turn model for adaptive routing. IEEE Transactions on Parallel Distributed Systems 11(7), 729–738 (2000)
11. Dally, W.J., Seitz, C.: The torus routing chip. Journal of Distributed Computing 1(3), 187–196 (1986)
12. Dally, W.J., Seitz, C.: Deadlock-free message routing in multiprocessor interconnection networks. IEEE Transactions on Computers C(36), 547–553 (1987)
13. Duato, J.: A new theory of deadlock-free adaptive routing in wormhole networks. IEEE Transactions on Parallel and Distributed Systems 4(12), 1320–1331 (1993)
14. Duato, J.: A necessary and sufficient condition for deadlock-free routing in wormhole networks. IEEE Transactions on Parallel and Distributed Systems 6(10), 1055–1067 (1995)
15. Duato, J., Lysne, O., Pang, R., Pinkston, T.M.: Part I: A theory for deadlock-free dynamic network reconfiguration. IEEE Transactions on Parallel and Distributed Systems 16(5), 412–427 (2005)
16. Fazzino, F., Palesi, M., Patti, D.: Noxim: Network-on-Chip simulator. http://noxim.sourceforge.net
17. Gindin, R., Cidon, I., Keidar, I.: NoC-based FPGA: architecture and routing. In: International Symposium on Networks-on-Chip, pp. 253–264. IEEE Computer Society (2007)
18. Glass, C.J., Ni, L.M.: The turn model for adaptive routing. Journal of the Association for Computing Machinery 41(5), 874–902 (1994)
19. Goossens, K., Bennebroek, M., Hur, J.Y., Wahlah, M.A.: Hardwired networks on chip in FPGAs to unify functional and configuration interconnects. In: IEEE International Symposium on Networks-on-Chip, pp. 45–54 (2008)
20. Guz, Z., Walter, I., Bolotin, E., Cidon, I., Ginosar, R., Kolodny, A.: Network delays and link capacities in application-specific wormhole NoCs. VLSI Design (2007)

21. Holsmark, R.: Deadlock free routing in mesh networks on chip with regions. Licentiate thesis, Linköping University, Department of Computer and Information Science, The Institute of Technology (2009)
22. Holsmark, R., Kumar, S.: Design issues and performance evaluation of mesh NoC with regions. In: IEEE Norchip, pp. 40–43. Oulu, Finland (2005)
23. Holsmark, R., Kumar, S.: Corrections to Chen and Chiu's fault tolerant routing algorithm for mesh networks. Journal of Information Science and Engineering **23**(6), 1649–1662 (2007)
24. Hu, J., Marculescu, R.: DyAD – smart routing for networks-on-chip. In: ACM/IEEE Design Automation Conference, pp. 260–263. San Diego, CA (2004)
25. Hu, J., Marculescu, R.: Energy- and performance-aware mapping for regular NoC architectures. IEEE Transactions on Computer-Aided Design of Integrated Circuits and Systems **24**(4), 551–562 (2005)
26. Jouraku, A., Koibuchi, M., Amano, H.: L-turn routing: An adaprive routing in irregular networks. Tech. Rep. 59, IEICE (2001)
27. Kim, J., Park, D., Theocharides, T., Vijaykrishnan, N., Das, C.R.: A low latency router supporting adaptivity for on-chip interconnects. In: ACM/IEEE Design Automation Conference, pp. 559–564 (2005)
28. Kumar, S., Jantsch, A., Soininen, J.P., Forsell, M., Millberg, M., Oberg, J., Tiensyrja, K., Hemani, A.: A network on chip architecture and design methodology. In: IEEE Computer Society Annual Symposium on VLSI, p. 117 (2002)
29. Lee, S.E., Bagherzadeh, N.: A high level power model for network-on-chip (NoC) router. Computers and Electrical Engineering **35**(6), 837–845 (2009)
30. Lotfi-Kamran, P., Daneshtalab, M., Lucas, C., Navabi, Z.: BARP-a dynamic routing protocol for balanced distribution of traffic in NoCs. In: ACM/IEEE Design Automation Conference, pp. 1408–1413 (2008)
31. Lysne, O., Pinkston, T.M., Duato, J.: Part II: A methodology for developing deadlock-free dynamic network reconfiguration processes. IEEE Transactions on Parallel and Distributed Systems **16**(5), 428–443 (2005)
32. Mejia, A., Flich, J., Duato, J., Reinemo, S.A., Skeie, T.: Segment-based routing: An efficient fault-tolerant routing algorithm for meshes and tori. In: International Parallel and Distributed Processing Symposium. Rhodos, Grece (2006)
33. Mejia, A., Palesi, M., Flich, J., Kumar, S., Lopez, P., Holsmark, R., Duato, J.: Region-based routing: A mechanism to support efficient routing algorithms in NoCs. IEEE Transactions on Very Large Scale Integration Systems **17**(3), 356–369 (2009)
34. Murali, S., Micheli, G.D.: Bandwidth-constrained mapping of cores onto NoC architectures. In: Design, Automation, and Test in Europe, pp. 896–901. IEEE Computer Society (2004)
35. Ni, L.M., McKinley, P.K.: A survey of wormhole routing techniques in direct networks. IEEE Computer **26**, 62–76 (1993)
36. Nilsson, E., Millberg, M., Oberg, J., Jantsch, A.: Load distribution with the proximity congestion awareness in a network on chip. In: Design, Automation and Test in Europe, pp. 1126–1127. Washington, DC (2003)
37. Palesi, M., Kumar, S., Holsmark, R.: A method for router table compression for application specific routing in mesh topology NoC architectures. In: SAMOS VI Workshop: Embedded Computer Systems: Architectures, Modeling, and Simulation, pp. 373–384. Samos, Greece (2006)
38. Palesi, M., Kumar, S., Holsmark, R., Catania, V.: Exploiting communication concurrency for efficient deadlock free routing in reconfigurable NoC platforms. In: IEEE International Parallel and Distributed Processing Symposium, pp. 1–8. Long Beach, CA (2007)
39. Palesi, M., Holsmark, R., Kumar, S., Catania, V.: Application specific routing algorithms for networks on chip. IEEE Transactions on Parallel and Distributed Systems **20**(3), 316–330 (2009)
40. Palesi, M., Kumar, S., Catania, V.: Bandwidth aware routing algorithms for networks-on-chip platforms. Computers and Digital Techniques, IET **3**(11), 413–429 (2009)

41. Pande, P.P., Grecu, C., Jones, M., Ivanov, A., Saleh, R.: Performance evaluation and design trade-offs for network-on-chip interconnect architectures. IEEE Transactions on Computers **54**(8), 1025–1040 (2005)
42. Sancho, J.C., Robles, A., Duato, J.: A flexible routing scheme for networks of workstations. In: International Symposium on High Performance Computing, pp. 260–267. London, UK (2000)
43. Schroeder, M.D., Birrell, A.D., Burrows, M., Murray, H., Needham, R.M., Rodeheffer, T.L., Satterthwaite, E.H., Thacker, C.P.: Autonet: a high-speed, self-configuring local area network using point-to-point links. IEEE Journal on Selected Areas in Communications **9**(8), 1318–1335 (1991)
44. Shang, L., Peh, L.S., Jha, N.K.: Dynamic voltage scaling with links for power optimization of interconnection networks. In: 9th International Symposium on High-Performance Computer Architecture, p. 91. IEEE Computer Society (2003)
45. Stensgaard, M.B., Sparsø, J.: ReNoC: A network-on-chip architecture with reconfigurable topology. In: IEEE International Symposium on Networks-on-Chip, pp. 55–64 (2008)
46. Upadhyay, J., Varavithya, V., Mohapatra, P.: A traffic-balanced adaptive wormhole routing scheme for two-dimensional meshes. IEEE Transactions on Computers **46**(2), 190–197 (1997)
47. Varatkar, G., Marculescu, R.: Traffic analysis for on-chip networks design of multimedia applications. In: ACM/IEEE Design Automation Conference, pp. 510–517 (2002)
48. Wolkotte, P.T., Smit, G.J., Kavaldjiev, N., Becker, J.E., Becker, J.: Energy model of networks-on-chip and a bus. In: International Symposium on System-on-Chip, pp. 82–85 (2005)
49. Wu, J., Jiang, Z.: Extended minimal routing in 2D meshes with faulty blocks. In: International Conference on Distributed Computing Systems, pp. 49–54 (2002)
50. Ye, T.T., Benini, L., Micheli, G.D.: Packetization and routing analysis of on-chip multiprocessor networks. Journal of System Architectures **50**(2–3), 81–104 (2004)

Chapter 6
Adaptive Data Compression for Low-Power On-Chip Networks

Yuho Jin, Ki Hwan Yum, and Eun Jung Kim

Abstract With the recent design shift toward increasing the number of processing elements in a chip, supports for low power, low latency, and high bandwidth in on-chip interconnect are essential. Much of the previous work has focused on router architectures and network topologies using wide/long channels. However, such solutions may result in a complicated router design and a high interconnect power/area cost. In this chapter, we present a method to exploit a table-based data compression technique, relying on value patterns in cache traffic. Compressing a large packet into a small one saves power consumption by reducing required operations in network components and decreases contention by increasing the effective bandwidth of shared resources. The main challenges are providing a scalable implementation of tables and minimizing the latency overhead of compression. We propose a shared table scheme that needs one encoding and one decoding tables for each processing element, and a management protocol that does not require in-order delivery. This scheme eliminates table size dependence on a network size, which realizes scalability and reduces overhead cost of table for compression. Our simulation results are presented for 8-core and 16-core designs. Overall, our compression method improves the packet latency up to 44% with an average of 36% and reduces the network power consumption by 36% on average in 16-core tiled design.

6.1 Introduction

With current technology scaling, future chip multiprocessors (CMPs) will accommodate many cores and large caches for high performance/energy-efficiency [12, 25, 29, 32]. This ever-increasing integration of components in a single chip has embarked the beginning of communication-centric designs [4, 8, 9, 17]. In this trend,

Y. Jin (✉)
Department of Electrical Engineering, University of Southern California,
3740 McClintock Ave., Los Angeles, CA 90089, USA
e-mail: yujin@usc.edu

C. Silvano et al. (eds.), *Low Power Networks-on-Chip*,
DOI 10.1007/978-1-4419-6911-8_6, © Springer Science+Business Media, LLC 2011

an on-chip network have been gaining wide acceptance for scalable communication architecture by providing connectivity between cores and other components. An on-chip network must provide high bandwidth, low latency, and low power consumption. Failing any of these objectives leads to a significantly detrimental impact on the overall system design.

Low power consumption support in an on-chip network is important, after power becomes a first-class constraint in computer system designs. In the Intel 80-core chip [12], a five-port router for the mesh network consumes a significant portion (28%) of the total tile power, which does not meet the chip power envelope (less than 10% of the total chip power). At the same time, an on-chip network should support low-latency and high-throughput that lead to high system performance, because a network is deeply integrated with on-chip caches and off-chip memory controllers. With these objectives, using intelligent routers [23, 24], high-radix topologies [16], or wide channels can achieve performance benefits but increase power/area costs that must be constrained within the budget of a subsystem. For the area constraint, conserving metal resources for channels can provide more space for silicon logic used for cores or caches [13]. To satisfy these stringent power and area budgets of a chip, simple router designs and network topologies are desirable. Therefore, rather than using complicated designs with over-provisioned resources, maximizing resource utilization in the existing design can achieve performance efficiency with constraint satisfaction.

Data compression has been adopted in hardware designs to achieve bandwidth and power savings, because the total amount of data for processing or storing can be reduced significantly. Cache compression increases the cache capacity by compressing recurring values and accommodating more blocks in a fixed space [1, 10]. Bus compression also expands the bus width by encoding the wide data as the small size code [2, 5]. Recently, data compression is explored in the on-chip network domain for performance and power [8].

In this research, we investigate adaptive data compression for on-chip network power/performance optimization and propose a cost-effective implementation. Our design uses a table-based compression approach by dynamically tracking value patterns in traffic. Using a table for compression hardware can process diverse value patterns adaptively rather than taking static patterns [8] to increase data compressibility. However, the table for compression requires non-scalable area cost to keep data patterns on a flow[1] basis. In other words, the number of tables depends on the network size, as communication cannot be globally managed in a switched network. To address this problem, we present the shared table scheme that can store identical values as a single entry across different flows. In addition, the management protocol for data consistency between an encoding table and a decoding table operates in a distributed way so that it allows out-of-order delivery in a network. In a suite of scientific and commercial multi-threaded benchmarks, we identify 69% compressibility of cache traffic on average for two CMP designs; eight-core with

---

[1] A flow represents a pair of source and destination. Therefore, an n-node network has n^2 flows.

static non-uniform cache architecture (SNUCA) [15] and 16-core tiled. Additionally, we show that a large portion of communication data is shared among different flows. The detailed simulation results reveal that data compression saves the energy by 36% (up to 56%) and improves the latency by 36% (up to 44%) in 16-core tiled-design.

This chapter is organized as follows. We briefly discuss related work in Sect. 6.2. We describe a table-based data compression in Sect. 6.3. We propose a scalable low-cost implementation of compression table hardware in Sect. 6.4. We describe evaluation methodologies in Sect. 6.5, present simulation results in Sect. 6.6, and conclude this chapter in Sect. 6.7.

6.2   Related Work

This research is motivated by a large body of prior work in value-centric architectures. Particularly, our work shares some common interests with cache compression and bus compression.

Value locality. Value locality as a small portion of recurred values by load instructions was reported in programs to predict load values [18]. Furthermore it is shown that programs have a set of frequent values across all load and store instructions [36].

Cache compression. Cache compression has been proposed to expand the cache capacity by packing more blocks than given by the space [1, 10]. In [1], the frequent pattern compression (FPC) scheme is used to store a variable number of blocks in the data array of the L2 cache. Due to the increased hit latency for decompression, they developed the adaptive scheme to determine whether a block is stored in a compressed form. Apart from compression hardware cost, cache compression requires significant modification to existing cache designs for flexible associativity management. In [33], the data array of L1 data cache is partitioned into encoded and unencoded parts. Accesses to the frequent values need only the small encoded data array, while the non-frequent values need serial accesses in both encoded data array and regular data array. This design aims to save energy at the cost of additional cycles needed to access nonfrequent values.

Bus compression. Bus compression can increase the bandwidth of a narrow bus for wide data. In [5], Bus-Expander stores the repeated high order bits of data into a table. For data transfer, an index into the table is sent along with the lower bits of the data. All the tables on the bus maintain the same content by snooping all the transferred values. However, a snooping operation needs broadcast support in a switched network and large bandwidth requirement. Also, a replacement in a table causes another replacement in all the tables, though newly placed data are directly relevant to only two tables in a sender and a receiver. This global replacement can evict a productive index for compression and result in a low compression rate. Power protocol [2] takes the similar scheme for bus energy reduction.

Bus-encoding techniques have been proposed to reduce the energy consumed in high-capacitance buses [20, 26, 31]. By detecting bit transition patterns on a bus, encoding hardware converts data into a low-transition form stored in a table. Introducing a special code for encoding further reduces energy consumption of the bus. An extra-bit line is needed to indicate whether the data are encoded or not. Bus-invert encoding transmits either original or inverted data depending on which would result in a small number of bit transitions [27]. Transition pattern encoding encodes data into a pre-built code that accounts for both inter-wire and intra-wire transitions [26]. A content-addressable memory (CAM)-based value table is commonly used to store repeated data and convert them into the energy-efficient encoded indices [2, 31, 34]. Liu et al. used the similar approach by adding a value cache to a bus for communication energy reduction in CMPs [19]. All the encoding techniques need an encoder at a sender side and a decoder at a receiver side to synchronize consistency between them.

Most of the above schemes assume bus-style interconnects, where data for compression are perfectly synchronized across all the nodes. A switched network, where each node needs to communicate with multiple nodes asynchronously, makes this problem challenging. Simply duplicating tables on a per-flow basis is not scalable toward large-scale networks in many-core processors. Moreover, compression tends to increase communication latency because compression process is performed before communication process. Thus, we need to develop a compression solution to minimize the negative impact on performance.

6.3 Data Compression In On-Chip Networks

In this section, we briefly present the on-chip network architecture and discuss benefits from data compression. Next, we propose a table-based data compression scheme to reduce the packet payload size.

6.3.1 On-Chip Network Architecture

Each processing element (PE) such as core, cache, and special processing engine is interconnected through a network. A switched network consists of routers and links whose connection determines a network topology.

Figure 6.1 shows two 8- and 16-core CMP systems with each network layout. The first design is SNUCA-CMP integrating many fast cache banks to reduce long access time of the monolithic L2 cache [3]. In an 8×8 mesh network, each router connects four cache banks and bridges to four neighboring routers.[2]

---

[2] Most routers have eight ports. Additionally, 12 routers have nine ports to connect a core (eight routers at periphery) or a memory controller (four routers in the center).

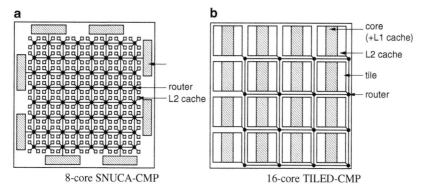

Fig. 6.1 On-chip networks in CMPs. (**a**) 8-Core SNUCA-CMP. (**b**) 16-Core TILED-CMP

The second design is TILED-CMP connecting homogeneous tiles to aim for the many-core paradigm. Each tile has a core, private L1 instruction/data caches, a part of a shared L2 cache, and a router. An N-core CMP has an N-tile network. Each router has two local ports for L1/L2 caches in its own tile and four ports to neighbor tiles for a mesh network.

In both designs, most communication is cache request/response messages and coherence messages for shared memory systems. Traffic has a bimodal-length distribution, depending on whether communication data includes a cache block or not. In other words, a packet payload has one of the following: address only (*address packet*), or both address and cache block (*data packet*).

Router. The router uses wormhole switching for small buffer cost, virtual channels (VCs) for head-of-line (HOL) blocking, and credit-based flow control for buffer overflow protection. The pipeline stages of a conventional router consist of route computation (RC), VC allocation (VA), switch allocation (SA), and switch traversal (ST) [7]. When a packet (a head flit of the packet) arrives at a router, the RC stage first directs a packet to a proper output port of the router by looking up a destination address. Next, the VA stage allocates one available VC of the downstream router determined by the RC stage. The SA stage arbitrates input and output ports of the crossbar, and then successfully granted flits traverse the crossbar in ST stage. The RC and VA stages are required only for head flits in this four-stage pipeline. In our study, we use a two-stage pipeline, which adopts lookahead routing and speculative switch allocation [24]. Lookahead routing removes the RC stage from the pipeline by making a routing decision one hop ahead of the current router. Speculative switch allocation enables the VA stage to be performed with the SA stage simultaneously.

Network interface. A network interface (NI) allows a PE to communicate over a network. An NI is responsible for packetization/de-packetization of data and flit[3]

---

[3] A flit can be subdivided into multiple physical transfer units (phits). Each phit is transferred across a channel in a single cycle. We assume that the phit size is equal to the flit size in this chapter.

fragmentation/assembly for flow control as well as high-level functions such as end-to-end congestion and transmission error control.

Link. Links for connecting routers are implemented as parallel global wires on metal resources. Setting a link width equal to the address packet size can increase link utilization and allow more metal resources for power and ground interconnects. Buffered wires are modeled to fit link delay within a single cycle [11].

6.3.2 Compression Support

In a switched network, communication data are transmitted as a packet. At a sender NI, a packet payload is split into multiple flits for flow control and then enters into a network serially. After traversing the network, all flits belonging to the same packet are concatenated and restored to the original packet.

If a certain value appears repeatedly in communication data, it can be transmitted as an encoded index, while any non-recurring value is transmitted in the original form. This is done by accessing a value encoding table that stores recurring values in the sender NI. When a packet that has encoded indices arrives, each index is restored to the original value by accessing the value decoding table in the receiver NI. Because the index size is much smaller than the value size, encoding can compress the packet. For correctness, we must ensure that a decoder successfully restores the original value associated with the encoded index.

Figure 6.2 shows an example of how to encode packet payload data as flits. We assume that the value size for a single encoding operation is the same as the flit size. In a compressed packet, the encoded indices (e2, e4, e5) follow the original data (v1, v3). This structure enables multiple flits to be successfully packed into a flit. Although it changes the data order of the original packet, it simplifies

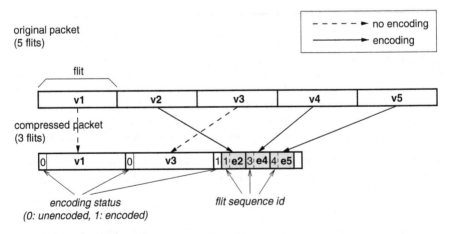

Fig. 6.2 Packet compression example

encoded index alignment with unencoded values and flit assembly at the receiver side. Because packet payload data can be partially compressed, reconstruction of the original packet requires two additional data: one bit indicating an encoding status for each flit and a flit sequence identifier for encoded flits to arrange all the flits in order of the original packet data. We do not consider specific energy-aware coding when building an index [20, 26, 31], because sharing links for different flows makes it hard to predict wire-switching activities.

We show how packet compression changes the packet delivery latency and power consumption. The contention-free packet delivery latency (T_0) consists of the router delay (T_r), wire delay (T_w), and serialization latency (T_s). The hop count (H) determines the total router delay as (HT_r) and the total length of wire that affects the wire delay (T_w). The serialization latency (T_s) is determined as L/b, where the packet length is L and the channel bandwidth is b.

$$T_0 = HT_r + T_w + T_s(= L/b). \tag{6.1}$$

As network load increases, more conflicts on ports and channels contribute to longer intra-router delay so the contention delay (T_c) is appended to T_0.

Compressing a long packet (L) into a short packet $(L' \leq L)$ reduces serialization latency but has extra encoding and decoding latencies $(T_e$ and $T_d)$.

$$T_0' = HT_r + T_w + T_s'(= L'/b) + T_e + T_d. \tag{6.2}$$

Encoding and decoding latencies are determined by the product of the encoding/decoding operation count and the unit operation latency. Compression may increase normal contention-free latency due to this latency overhead. However, in wormhole switching, reduced packet size can achieve better resource utilization. Because average load for each router is reduced, this leads to less contention for shared resources in a router.

Energy consumption of a packet (E_p) is given by

$$E_p = L/b(DE_{link} + HE_{router}), \tag{6.3}$$

where D is Manhattan distance, H is hop count, E_{link} is unit length link energy consumption, and E_{router} is router energy consumption. By reducing L to L', compression reduces the number of flits in a packet from L/b to L'/b. Hence router and link energy for a packet can be reduced at the cost of encoder and decoder energy $(E_{enc}$ and $E_{dec})$. Energy consumption of a packet with compression support can be derived as:

$$E_p' = L'/b(DE_{link} + HE_{router}) + E_{enc} + E_{dec}. \tag{6.4}$$

Longer-distance communication packets (larger D and H) can save more energy, because the required energy for routers and links becomes much larger than the energy for compression. This additional energy from compression/decompression operations mainly depends on the size of value tables. Next, we explain organization of table that stores recurring values.

6.3.3 Table Organization

In an n-PE network, each PE needs n encoding tables to convert a value into an index and n decoding tables to recover a value from a received index. We call this organization *private table* scheme, because a table needs to be maintained separately for each flow. An encoding table that has value-index entries is constructed using a CAM-tag cache, where a value is stored in a tag array for matching while an associated encoded index is stored in a data array. Those indices can be pre-built or read-only because they do not need to be altered at runtime. In a decoding table that has index-value entries, a received index is decoded to select an associated value. Since a decoding table is simply organized as a direct-mapped cache, a received index can uniquely identify one value.

One encoding table and its corresponding decoding table need to be consistent to correctly recover a value from an encoded index. Both tables should have the same number of entries and use the same replacement policy. If a packet data causes a replacement in the encoding table, it must also replace the same value in the decoding table upon arrival. Furthermore, a network must provide flow-based in-order packet delivery to make replacement actions for both tables in the same order. To guarantee in-order delivery, a network needs a large reorder buffer at receivers, which requires additional area cost, or it should restrict dynamic management such as adaptive routing. We use a deterministic routing algorithm and a first-come first-serve (FCFS) scheduler in arbiters.

The private table scheme relies on decoding ability from per-flow value management. This does not provide a scalable solution as the network size increases. A substantial chip area must be dedicated for implementing private tables. Moreover, it is possible that identical values are duplicated across different tables, because each table is exclusively used for a single flow. Therefore, despite the large table capacity, the private table scheme cannot manage many distinct values effectively.

6.4 Optimizing Compression

To overcome a huge implementation cost of the private table scheme, we present another table organization scheme and its management in this section. We call this *shared table* scheme, as each PE has single encoding/decoding tables shared by all packets from/to that PE.

6.4.1 Shared Table Structure

Each PE has one encoding table and one decoding table by merging the same values across different flows. Value analysis in two CMP architectures reveals that one sender transmits same values to a large portion of receivers and vice versa (see the detailed results in Sect. 6.6.1). Therefore, having a network-wide single entry for each value in tables can dramatically reduce the table size. Unlike the private table scheme, a receiver finds value patterns used for encoding. When a receiver places a new value in the decoding table, it notifies the corresponding sender of the new value and the associated index. After a sender receives the index for the new value, it can begin to compress that value. Thus, the index part in the decoding table is read-only but the elements of the index vector in the encoding table can be updated.

In an encoding table, a value in each entry is associated with multiple indices constructed as a vector. A position of the index vector indicates one PE as the receiver. Each element has an index value that will select one entry in the corresponding decoding table. In a decoding table, one entry has three fields: a value, an index, and a use-bit vector. Each bit in the use-bit vector tells if the corresponding sender transmits the associated value as index. Hence, the number of columns related to the index vector part in the encoding table and to the use-bit vector of the decoding table are equal to the number of PEs. Figure 6.3 presents the table structure for 16-PE network, showing as an example a possible encoding and decoding table for PE4 and PE8, respectively. The encoding table shows that **A**, the value of the first entry, is used by six receiver PEs (0, 4, 7, 8, 12, 14). Likewise, the decoding table shows that **A**, the value of the second entry, is used by four sender PEs (2, 4, 11, 15). The encoding table in PE4 indicates that three values (**A**, **B**, and **D**) can be encoded when transmitted to PE8.

a

value	index vector
A	00 · · · · 00 · · · 01 01 · · · 00 · 00
B	11 10 · 00 · 01 · 00 11 · 11 10 · 01 10 11
C	· 11 · · · 10 · · 01 01 · · · 10
D	10 · 01 · · 00 · 10 · 00 11 · · ·

for PE8

Each element shows a binary index for value at decoder.

Encoding Table in PE4

b

value	index	use–bit vector
E	00	1 1 0 0 0 1 1 0 1 0 0 0 0 1 0 0
A	01	0 0 1 0 1 0 0 0 0 0 1 0 0 0 1
D	10	1 0 0 0 1 0 0 0 0 1 1 0 0 1 0 0
B	11	0 1 0 0 1 0 0 0 1 0 0 0 0 1 0 1

for PE4

Each bit indicates a value status of encoder.

Decoding Table in PE8

Fig. 6.3 Shared table structure. (**a**) Encoding table in PE4. (**b**) Decoding table in PE8

6.4.2 Shared Table Consistency Management

Value-index association between a sender and a receiver must be consistent for correct compression/decompression. A sender must not transmit an index from which a receiver cannot restore a correct value in its decoding table. Specifically, an index associated with the value in the decoding table can be used in multiple encoding tables. Changing a value associated with an index in the decoding table requires some actions in encoding tables that use the index. Thus, a consistency mechanism is required between encoding tables and decoding tables.

For this purpose, we propose the table management protocol for the shared table scheme. In this protocol, a receiver tracks new values while a sender tracks new values in the private table scheme. As a result, inserting a new value into the decoding table starts at a receiver. When a specific value appears repeatedly, a receiver does one of the following two operations: (a) if the new value is not found in the decoding table, the receiver *replaces* an existing value with the new value. (b) if the value is found but a use-bit for the sender is not set, the receiver *updates* the corresponding use-bit of the decoding table. After either replacement or update operation, the receiver notifies the corresponding sender of the associated index for a new value. Finally, the sender inserts the index and the new value in the encoding table.

Figure 6.4a illustrates a replacement example with two encoding tables (EN0 for PE0, EN15 for PE15) and one decoding table (DE1 for PE1) in a 16-PE network. DE1 has two values (**A** and **B**) for EN0 and three values (**A**, **B**, and **G**) for EN15. When the new value **F** comes to DE1 (①), the decoding table needs a replacement for **F** and decides to evict **B**. Then, it requests all the related encoding tables (②) for invalidation of **B** (③) and waits for invalidation acknowledgment from the encoding tables (④). DE1 replaces the old value **B** with the new value **F** (⑤) and then sends replacement to related encoding tables (⑥ and ⑦).

Figure 6.4b illustrates an update example. The sender (EN0) transmits the new value **G**, which is in the decoding table, but the use-bit for EN0 is not set. DE1 sets the corresponding bit of the use-bit vector (②) and sends UPDATE command for **G** to EN0 (③). Finally, EN0 has **G** (④).

This management protocol makes sure that the values that the decoding table has are convertible into indices at the encoding table. Note that an encoding table update operation for a sender is initiated by a receiver. A decoding table can have more entries than an encoding table to accommodate more distinct values from different senders.

6.4.3 Increasing Compression Effectiveness

Because a single decoding table handles value locality from multiple flows, the shared table scheme may experience many replacement operations due to the increased number of non-recurring values, causing a low compression rate. One replacement operation in a decoding table requires at least two packet transmissions to be consistent with an encoding table, increasing control traffic.

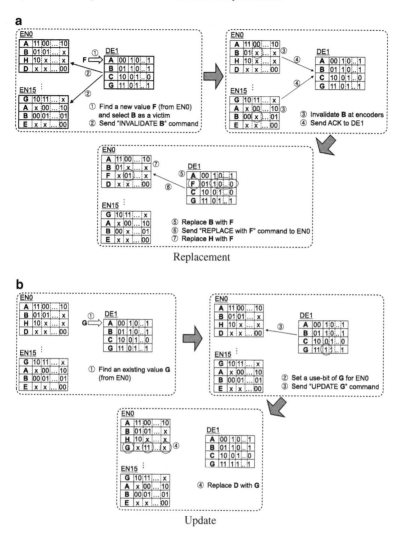

Fig. 6.4 Shared table management. (**a**) Replacement. (**b**) Update

To mitigate this problem, we use another table, value locality buffer (VLB), that filters undesirable replacement for the decoding table. VLB has a simple associative structure where each entry has a value and a counter for frequency. When a value arrives at a receiver, it is stored in VLB. VLB follows the least frequently used replacement. Whenever a value hit occurs in VLB, the counter of the entry is increased by one. When a counter is saturated, the associated value results in replacement for a decoding table. In other words, VLB is used for confirming temporal locality of new values.

Performance improvement techniques that reduces encoding/decoding latency from compression hardware can be developed by overlapping encoding with flit injection and dynamically controlling compression for workload [14].

6.5 Methodology

Our evaluation methodology consists of two parts: first, we used Simics [21] full-system simulator configured for UltraSPARCIII+ multiprocessors running Solaris 9 and GEMS [22] that models directory-based cache coherence protocols to obtain real workload traces. Second, we evaluated the performance and estimated the dynamic power consumption of varying compression schemes using an interconnection network simulator that models routers, links, and NIs in detail.

Table 6.1 shows main parameters of 8-core SNUCA-CMP and 16-core TILED-CMP designs for 45 nm technology as shown in Fig. 6.1. We chose the 4 GHz frequency, which respects power limitation in future CMPs guided by [17]. All the cache-related delay and area parameters are determined by Cacti [28]. Both designs accommodate 16 MB networked L2 cache and superscalar cores configured similarly with [3]. Assuming that each chip area is 400 mm^2, the cache network in SNUCA-CMP has 2 mm hop-to-hop links to connect 1 mm^2 banks and then the partitioned L2 cache lies on 256 mm^2. Tiles in a 4×4 mesh network of TILED-CMP are connected with 5 mm long links.

The network simulator that replays packets generated by the CMP simulator models the detailed timing and power behaviors of the routers, links, encoders, and decoders. The router in each CMP design is configured to fit its pipeline delay[4] for one cycle of 4 GHz clock frequency using the logical effort model [24]. Each router has four-flit buffers for each virtual channel and one flit contains 8B data. We estimate router power consumption from Orion [30]. High connectivity networks need high radix routers that require a longer pipeline delay and consume more power

Table 6.1 CMP system parameters for data compression

CMP design	SNUCA-CMP	TILED-CMP
Clock frequency	4 GHz	4 GHz
Core count	8	16
L1 I & D cache	2-way, 32 KB, 2 cycles	2-way, 32 KB, 2 cycles
L2 cache	16-way, 256×64 KB, 3 cycles (per bank)	16-way, 16×1 MB, 10 cycles (per bank)
L1/L2 cache block	64B	64B
Memory	260 cycles, 4 GB DRAM	260 cycles, 8 GB DRAM
coherence protocol	MOSI	MSI
network topology	8×8 mesh	4×4 mesh

---

[4] VC allocator ($R \rightarrow p$) has the longest latency among others and determines the delay of a router pipeline in both designs. Therefore, the number of VCs is selected for one clock cycle time.

Table 6.2 Delay and power characteristics of interconnect

(a) *Router*

CMP	Channel (p, v)	Delay (ns)	Buffer (pJ)	Crossbar (pJ)	Arbiter (pJ)	Leakage (pJ)
TILED	6, 3	0.250	11.48	34.94	0.22	9.05
SNUCA	8, 2	0.230	15.30	61.23	0.32	15.12
	9, 2	0.235	17.22	77.12	0.39	18.73

(b) *Link*

Delay	Dynamic power		Leakage power
	Wire-substrate	Inter-wire	One wire
183 (ps/mm)	1.135 (mW/mm)	0.634 (mW/mm)	0.0016 (mW/mm)

than low radix routers. Thus, the higher radix routers in SNUCA-CMP require more power consumption than those in TILED-CMP. Table 6.2a shows the router pipeline delay and the energy consumption of each router component. The second column specifies the channel property as the number of physical channels (p) and the number of VCs (v).

To overcome long global wire delay, repeaters are inserted to partition a wire into smaller segments, thereby making delay linear with its length. Because the link power behavior is sensitive to value patterns, we consider the actual bit pattern crossing a link and the coupling effect on adjacent wires [26]. We divide the wire capacitance (c_w) into two parts: wire-substrate capacitance (c_s) and inter-wire capacitance (c_i). c_i is known to become more dominant than c_s as technology shrinks [31]. Switching activities for both capacitances are computed from the data sequence traversing the link. For the two-wire link example, transition from 01 to 10 results in two wire-substrate and two inter-wire switching activities. Thus, we can drive energy drawn in a multiwire link (E_{link}).

$$E_{link} = 0.5 V_{dd}^2 \left(\alpha \left(\frac{k_{opt}}{h_{opt}} (c_0 + c_p) + c_s \right) + \beta c_i \right) L, \qquad (6.5)$$

where α and β are the transition counts for wire-substrate and inter-wire, respectively. At 45 nm targeting year 2010, global wires having 135 nm pitch has 198 fF mm^{-1}, where inter-wire capacitance is four times higher than wire-substrate capacitance. Table 6.2b shows the delay and power models of the global wire.

The benchmarks considered in this research are six parallel scientific (SPEComp) and two server (SPECjbb2000, SPECweb99) workloads. SPEComp programs are compiled on a Sun Studio 11 compiler and executed for parallel regions with reference data sets.

6.6   Experimental Results

We conducted experiments to examine how communication compression affects the performance and power consumption of on-chip interconnection networks. In 8B-wide channel networks for 64B cache block data transportation, we assume that the address packet has a single flit and the data packet is broken down to nine flits, where the first flit has an address and other flits have the part of cache block data starting from its most significant bit position. We apply compression only to cache block data, because address compression requires another table and does not give a high return for packet length reduction. Four 2B-entry tables are used to compress the corresponding part of 8B flit data concurrently.

6.6.1 Compressibility and Value Pattern

We examine the value compressibility of cache traffic in CMPs. Since one cache block contains 16 4B words, data redundancy can exist within a cache block. In addition, a value pattern detection method such as LRU and LFU affects compressibility. LRU replacement gives significance to the recently used one, while LFU replacement runs based on the reuse frequency. We put two fixed-size tables at both sender and receiver sides like private table and use a hit rate as a compressibility metric. We change the table size by varying the number of entries from 4 to 256 and the size of entry from 1 to 64B.

Figure 6.5 shows a trend of the average hit rates for two replacement policies[5]. As the size of entry is smaller or the number of entries is larger, the hit rate is better. For a fixed size table, making the entry size smaller increases the hit rate better than providing more number of entries owing to the partial value redundancy. In 128B-table with LFU in TILED-CMP, $2B \times 64$ has 5, 13, and 30% higher hit rate than $4B \times 32$, $8B \times 16$, and $16B \times 8$, respectively. Although it is not easy to show which replacement policy is better in our experiments, LFU hit rate is less sensitive across the different number of entries in the table, which implies that multi-threaded programs have a set of frequent values like single-threaded programs [36]. This result shows high compressibility in cache traffic even with small tables.

Figure 6.6 compares compression ratios for three compression techniques: frequent pattern compression (FPC), our table-based approach, and LZW-variant compression algorithm (denoted as gzip in figures). FPC compression algorithm detects six different sizes of zero data and one word of repeated bytes, which needs three-bit prefix compression overheads for each word. We use the gzip unix utility for LZW compression that combines the LZW deflate algorithm with Huffman encoding of codewords in the dictionary. Our table approach uses eight-entry tables, which needs a three-bit index for compressed data. Additional overhead due to compression is one-bit compression status for each word in our scheme. While our scheme and FPC achieves fast compression, our table-based compression provides 16% higher compression ratios than FPC. Compared to the idealistic compression algorithm, the table-based approach achieves 56% of the gzip compression.

---

[5] We put the value table at the router rather than each node in SNUCA-CMP.

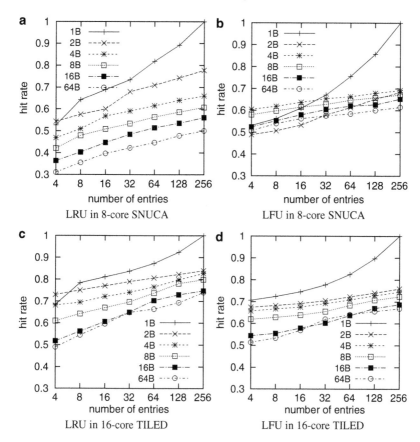

Fig. 6.5 Communication data compressibility with different replacement policies, entry sizes, and entry counts. (**a**) LRU in 8-core SNUCA. (**b**) LFU in 8-core SNUCA. (**c**) LRU in 16-core TILED. (**d**) LFU in 16-core TILED

We examine value sharing property by analyzing values across different flows that have a common source (sender) or destination (receiver). The destination sharing degree is defined as the average number of destinations per value. For a 10 K-cycle interval, we calculate the destination sharing degree for one source by taking the average number of destinations of each value, weighting it with the percentage of accesses that each value accounts for, and summing up the weighted destination counts. We finally take the average for all sources. Similarly, we obtain the source sharing degree. We do the same analysis considering only top n values ordered by the number of accesses. Due to the intractable number of values, we reset analysis results every 100 K cycles.

Figure 6.7 shows sharing degrees for 2B value in each benchmark. Regarding top (frequently accessed) four values shows much higher sharing degree than taking

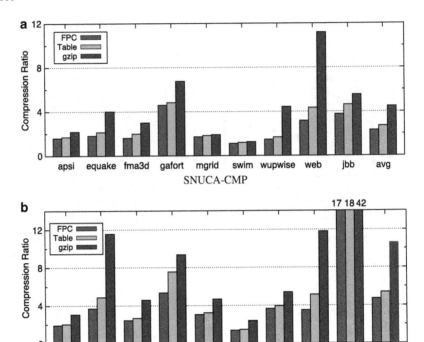

Fig. 6.6 Compression technique comparison. (**a**) SNUCA-CMP. (**b**) TILED-CMP

all values[6]. Particularly, TILED-CMP shows that top four values are used at almost 12 nodes (75% in the network) on average. This result suggests that organizing encoding/decoding tables by sharing frequent values can keep a high compression rate. We select 2B × 8 table and LFU policy, which is fairly small but has a high hit rate, to evaluate our compression techniques further.

6.6.2 Effect on Power Consumption

Figure 6.8 shows energy reduction relative to the baseline. The private and shared table schemes are indicated by **Pv** and **Sh**. A shared table has an eight-entry VLB in each decoding table. Private table (second bar) and shared table (third bar) save energy over the baseline (first bar) by 11.9% (TILED-CMP) and 12.2% (SNUCA-CMP) on average, respectively. As long-distance communication data is more involved with compression, energy saving becomes more effective. For example, *fma3d* exhibits the hop count as 3.16 and 42% energy saving for private table in

---

[6] The destination sharing degree (dest) is the average number of destinations for each value. The source sharing degree (src) is the average number of sources for each value.

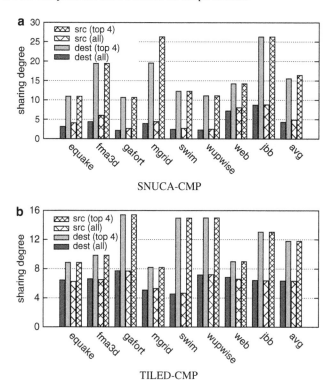

Fig. 6.7 Value spatial distribution. (**a**) SNUCA-CMP. (**b**) TILED-CMP

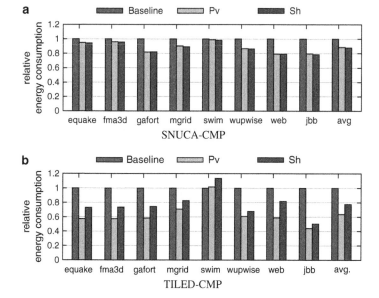

Fig. 6.8 Energy consumption comparison. (**a**) SNUCA-CMP. (**b**) TILED-CMP

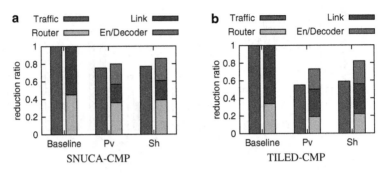

Fig. 6.9 Traffic and energy relationship. (**a**) SNUCA-CMP. (**b**) TILED-CMP

TILED-CMP, while *mgrid* exhibits the hop count as 2.76 and 28% energy saving. Our examination in *swim* for very low energy savings exhibits that *swim* has a huge number of different values and hence leads to a low hit rate on tables.

In link power estimation, we find that using link utilization overestimates its energy consumption rather than accounting bit patterns. In benchmarks, we observed that, link utilization shows an average of 11% (up to 24%), while bit pattern analysis for intra-wire switching gives us an average of 2% activity factor (up to 7%).

Figure 6.9 shows a traffic (left bar) and energy (right bar) relationship in different schemes. It further breaks down energy consumption in routers, links, and encoding/decoding tables. We observe that energy reduction is less than traffic reduction, because more repeated values in traffic implies smaller switching activity. Additionally, encoding indices from compression introduce a new pattern that is not in the original workload, causing extra switching activities. Routers consume 34% (TILED-CMP) and 39% (SNUCA-CMP) of the total network energy in baseline configurations. Note that SNUCA-CMP has higher-radix and more routers than TILED-CMP. In fact, energy consumption ratio for each component depends on network parameters (buffer depth, router radix, link length) and workload characteristics (average hop count, bit pattern). For the shared table scheme in TILED-CMP, encoders and decoders consume 25% of the total energy consumption.

The shared table achieves almost the same compression rate as the private table. The main reason is 50% in TILED-CMP and 59% SNUCA-CMP from high value sharing as explained in Sect. 6.6.1. The average hit rate in encoding tables of the shared table decreases by 6.4 and 1.8% each. The hit rates are listed in Table 6.3. We find that management traffic for the shared table scheme increases the overall traffic by less than 1%.

6.6.3 Effect on Packet Latency

Data compression introduces additional compression/decompression latency to the normal architecture. This overhead increases zero latency of packet delivery, as both

Table 6.3 Encoding table hit rates for private and shared tables

CMP design	Scheme	*equake*	*fma3d*	*gafort*	*mgrid*	*swim*	*wupwise*	*web*	*jbb*	avg.
SNUCA	Private	0.358	0.252	0.800	0.455	0.093	0.449	0.583	0.661	0.457
	Shared	0.355	0.251	0.792	0.453	0.094	0.447	0.532	0.661	0.448
TILED	Private	0.640	0.617	0.795	0.667	0.258	0.741	0.601	0.964	0.660
	Shared	0.521	0.525	0.781	0.662	0.253	0.727	0.519	0.955	0.618

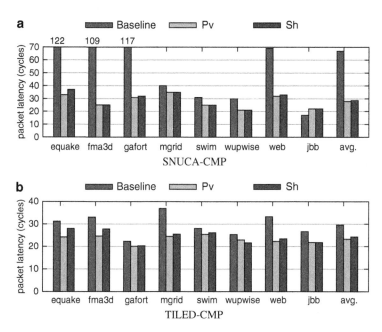

Fig. 6.10 Packet latency comparison. (**a**) SNUCA-CMP. (**b**) TILED-CMP

compression and decompression operations lie on the critical path of packet transmission. However, reducing the size of packets, in particular, converting a multi-flit packet into a single-flit packet (or with fewer flits) through data compression can significantly reduce resource contention in network architecture by decreasing injection load. Hence, virtual channel or switch allocation can proceed without any waiting delay from failed operations in previous cycles.

Figure 6.10 shows the average packet latency of different compression architectures compared to the baseline. In most benchmarks, we can see that private table (second bar) achieves latency reduction by 51 and 60% in TILED-CMP and SNUCA-CMP, respectively. A main factor for latency reduction is that transmitting compressed packets cancel out most contention or congestion behavior in our examined benchmarks. In particular, compression improves the latency dramatically by resolving high congestion in some benchmarks (*equake/fma3d/gafort/web* in TILED-CMP and *mgrid/web* in SNUCA-CMP).

The shared table penalizes the latency by only 4.7% (TILED-CMP) and 0.2% (SNUCA-CMP) over private table (second bar). This penalty comes from decreased compressibility due to a hit loss in encoding tables. This small loss of compressibility can be traded off the low implementation cost.

Investigation of runtime latency behavior in the compression architecture compared to the baseline architecture reveals that sending compressed packets increases packet delivery latency slightly in low load due to compression latency. Additionally, for some time intervals, a set of sources instantly send large number of packets and hog interconnect bandwidth, thus leading to congestion in NIs and routers. This destructive behavior causes a latency increase tremendously by increasing queuing burden at NIs and resource contention at routers.

We observe low load in a network during most of the execution time. However, the time when congestion occurs determines average packet latency, because a large number of packets with high latency are transmitted at that time. We think that performance benefit of the system through data compression would be reduced in feedback simulation that models interaction in an on-chip network with other architectural components. This average low-load behavior motivates us for embedding the on-demand compression control techniques to enable compression only if compression is beneficial in terms of packet latency [14].

6.6.4 Compression Table Area Analysis

Value table hardware is a critical component that enables data compression in an on-chip interconnect. The large table can accommodate many different values in a network but incurs an increase in the area cost and access latency. Moreover, an on-chip network design is constrained by power and area budgets. This extra hardware cost should not inflate the design cost much.

To analyze the area cost of value table implementation, we estimate the table area as the total area of CAM and RAM cells. The associative search part of a table are implemented as CAM and other parts are constructed as RAM. Table 6.4 shows the required number of RAM and CAM cells for each scheme, where n is the PE count, v is the value size in bits, and e/d is the number of entries in encoding/decoding table. We do not account for counters and valid bits.

Table 6.4 Value table area cost

	Private table		Shared table	
Encoder	Value:	$v \cdot e \cdot n$ (CAM)	Value:	$v \cdot e$ (CAM)
	Index:	$\log e \cdot e \cdot n$ (RAM)	Index vector:	$\log d \cdot e \cdot n$ (RAM)
Decoder	Index:	0	Index:	$\log d \cdot d$ (CAM)
	Value:	$v \cdot d \cdot n$ (RAM)	Value:	$v \cdot d$ (CAM)
			Use-bit vector:	$d \cdot n$ (RAM)
			VLB:	$v \cdot d$ (CAM)

Fig. 6.11 Area comparison
for private and shared tables

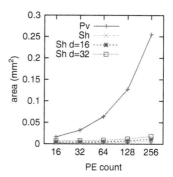

To present the trend of the table cost according to increasing number of PEs, we obtain the cell areas from [35]. Scaling these values at 45 nm technology gives us $0.638\,\mu m^2$ for RAM (six transistors) and $1.277\,\mu m^2$ for CAM (nine transistors). Figure 6.11 shows huge area overhead of the private table scheme (**Pv**) compared to the shared table scheme (**Sh**) with eight 8B entries at large PE count. In the shared table scheme, a decoding table can have more entries than an encoding table to increase compression rate. For the decoding tables with 16 (**Sh d=16**) and 32 (**Sh d=32**) entries of 8B value. Figure 6.11 illustrates that the increased area for the shared table scheme is still scalable.

6.6.5 Comparison with Wide/Long-Channel Networks

To verify the compression efficiency of the mesh network with moderately sized (8B) channel width compared to other wide/long-channel networks, we apply the shared table compression scheme with the latency hiding techniques [14] to a mesh network with 16B-wide links and an express cube [6] with one-hop express links. In the express cube, we use a deterministic routing algorithm that first uses the express channels and then regular channels.

Figure 6.12 shows energy consumption, packet latency, area, and overall efficiency as a product of three metrics. Left (w/o) and right (w/) groups represent the baseline and compression architectures each. Compression improves the latency on average by 43% in all networks. It should be mentioned that using deterministic routing in the express cube does not fully use its increased path diversity. As dynamic energy contributes a small portion of the total energy in our environment, compression does not achieve much energy savings. Moreover, adopting wide and long channels increases static energy consumption and takes more area in a chip. In summary, comparison results of compression in three networks show that the 8B-wide link mesh has overall efficiency higher than the 16B-wide link mesh by 11%, and the express cube by 59%.

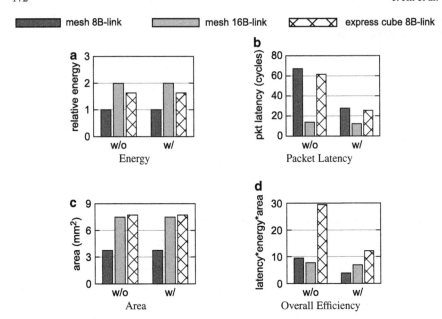

Fig. 6.12 Network comparison for TILED-CMP. (**a**) Energy. (**b**) Packet latency. (**c**) Area. (**d**) Overall efficiency

6.7 Conclusion

The current technology scaling trends indicate that the on-chip interconnect consumes a significant portion of power and area in future many-core systems. We introduce a table-based data compression framework that achieves power savings and bandwidth increasings of the existing on-chip network by adaptively collecting data patterns in traffic. Furthermore, compression enables the network to enjoy low-latency packet delivery at high load by reducing resource contention.

In this chapter, we present the optimization techniques to overcome the table cost for data compression implementation. We propose the shared table scheme, which stores identical values into a single entry from different sources or destinations and removes the network-size dependence for scalability. We also present the efficient table management protocol for consistency. Our simulation results in the TILED-CMP and SNUCA-CMP designs show that compression using the shared tables saves the energy consumption 12 and 12% and improves the packet latency by 50 and 59%, respectively.

Acknowledgements This work was supported in part by NSF grants CCF-0541360 and CCF-0541384. Yuho Jin is currently supported by the National Science Foundation under Grant no. 0937060 to the Computing Research Association for the CIFellows Project.

References

1. Alameldeen, A.R., Wood, D.A.: Adaptive Cache Compression for High-Performance Processors. In: Proceedings of ISCA, pp. 212–223 (2004)
2. Basu, K., Choudhary, A.N., Pisharath, J., Kandemir, M.T.: Power Protocol: Reducing Power Dissipation on Off-Chip Data Buses. In: Proceedings of MICRO, pp. 345–355 (2002)
3. Beckmann, B.M., Wood, D.A.: Managing Wire Delay in Large Chip-Multiprocessor Caches. In: Proceedings of MICRO, pp. 319–330 (2004)
4. Cheng, L., Muralimanohar, N., Ramani, K., Balasubramonian, R., Carter, J.B.: Interconnect-Aware Coherence Protocols for Chip Multiprocessors. In: Proceedings of ISCA, pp. 339–351 (2006)
5. Citron, D., Rudolph, L.: Creating a Wider Bus Using Caching Techniques. In: Proceedings of HPCA, pp. 90–99 (1995)
6. Dally, W.J.: Express Cubes: Improving the Performance of k-Ary n-Cube Interconnection Networks. IEEE Transactions on Computers **40**(9), 1016–1023 (1991)
7. Dally, W.J., Towles, B.: Principles and Practices of Interconnection Networks. Morgan Kaufmann, San Francisco (2003)
8. Das, R., Mishra, A.K., Nicopolous, C., Park, D., Narayan, V., Iyer, R., Yousif, M.S., Das, C.R.: Performance and Power Optimization through Data Compression in Network-on-Chip Architectures. In: Proceedings of HPCA, pp. 215–225 (2008)
9. Eisley, N., Peh, L.S., Shang, L.: In-Network Cache Coherence. In: Proceedings of MICRO, pp. 321–332 (2006)
10. Hallnor, E.G., Reinhardt, S.K.: A Unified Compressed Memory Hierarchy. In: Proceedings of HPCA, pp. 201–212 (2005)
11. Ho, R., Mai, K., Horowitz, M.: The Future of Wires. In: Proceedings of the IEEE, pp. 490–504 (2001)
12. Hoskote, Y., Vangal, S., Singh, A., Borkar, N., Borkar, S.: A 5-GHz Mesh Interconnect for a Teraflops Processor. IEEE Micro **27**(5), 51–61 (2007)
13. Jayasimha, D.N., Zafar, B., Hoskote, Y.: Interconnection Networks: Why They are Different and How to Compare Them. Tech. rep., Microprocessor Technology Lab, Corporate Technology Group, Intel Corp (2007). http://blogs.intel.com/research/terascale/ODI_why-different.pdf
14. Jin, Y., Yum, K.H., Kim, E.J.: Adaptive Data Compression for High-Performance Low-Power On-Chip Networks. In: Proceedings of MICRO, pp. 354–363 (2008)
15. Kim, C., Burger, D., Keckler, S.W.: An Adaptive, Non-Uniform Cache Structure for Wire-Delay Dominated On-Chip Caches. In: Proceedings of ASPLOS, pp. 211–222 (2002)
16. Kim, J., Balfour, J., Dally, W.J.: Flattened Butterfly Topology for On-Chip Networks. In: Proceedings of MICRO, pp. 172–182 (2007)
17. Kirman, N., Kirman, M., Dokania, R.K., Martínez, J.F., Apsel, A.B., Watkins, M.A., Albonesi, D.H.: Leveraging Optical Technology in Future Bus-based Chip Multiprocessors. In: Proceedings of MICRO, pp. 492–503 (2006)
18. Lipasti, M.H., Wilkerson, C.B., Shen, J.P.: Value Locality and Load Value Prediction. In: Proceedings of ASPLOS, pp. 138–147 (1996)
19. Liu, C., Sivasubramaniam, A., Kandermir, M.: Optimizing Bus Energy Consumption of On-Chip Multiprocessors Using Frequent Values. Journal of Systems Architecture **52**, 129–142 (2006)
20. Lv, T., Henkel, J., Lekatsas, H., Wolf, W.: A Dictionary-Based En/Decoding Scheme for Low-Power Data Buses. IEEE Transactions on VLSI Systems **11**(5), 943–951 (2003)
21. Magnusson, P.S., Christensson, M., Eskilson, J., Forsgren, D., Hållberg, G., Högberg, J., Larsson, F., Moestedt, A., Werner, B.: Simics: A Full System Simulation Platform. IEEE Computer **35**(2), 50–58 (2002)
22. Martin, M.M., Sorin, D.J., Beckmann, B.M., Marty, M.R., Xu, M., Alameldeen, A.R., Moore, K.E., Hill, M.D., Wood, D.A.: Multifacet's General Execution-driven Multiprocessor Simulator (GEMS) Toolset. Computer Architecture News **33**(4), 92–99 (2005)

23. Mullins, R.D., West, A., Moore, S.W.: Low-Latency Virtual-Channel Routers for On-Chip Networks. In: Proceedings of ISCA, pp. 188–197 (2004)
24. Peh, L.S., Dally, W.J.: A Delay Model and Speculative Architecture for Pipelined Routers. In: Proceedings of HPCA, pp. 255–266 (2001)
25. Sankaralingam, K., Nagarajan, R., Liu, H., Kim, C., Huh, J., Burger, D., Keckler, S.W., Moore, C.R.: Exploiting ILP, TLP, and DLP with the Polymorphous TRIPS Architecture. In: Proceedings of ISCA, pp. 422–433 (2003)
26. Sotiriadis, P.P., Chandrakasan, A.: Bus Energy Minimization by Transition Pattern Coding (TPC) in Deep Submicron Technologies. In: Proceedings of ICCAD, pp. 322–327 (2000)
27. Stan, M., Burleson, W.: Bus-Invert Coding for Low-Power I/O. IEEE Transaction on VLSI 3(1), 49–58 (1995)
28. Tarjan, D., Thoziyoor, S., Jouppi, N.P.: Cacti 4.0. Tech. Rep. HPL-2006-86, HP Laboratories (2006)
29. Taylor, M.B., Lee, W., Amarasinghe, S.P., Agarwal, A.: Scalar Operand Networks: On-Chip Interconnect for ILP in Partitioned Architecture. In: Proceedings of HPCA, pp. 341–353 (2003)
30. Wang, H., Zhu, X., Peh, L.S., Malik, S.: Orion: a Power-Performance Simulator for Interconnection Networks. In: Proceedings of MICRO, pp. 294–305 (2002)
31. Wen, V., Whitney, M., Patel, Y., Kubiatowicz, J.: Exploiting Prediction to Reduce Power on Buses. In: Proceedings of HPCA, pp. 2–13 (2004)
32. Wentzlaff, D., Griffin, P., Hoffmann, H., Bao, L., Edwards, B., Ramey, C., Mattina, M., Miao, C.C., III, J.F.B., Agarwal, A.: On-Chip Interconnection Architecture of the Tile Processor. IEEE Micro 27(5), 15–31 (2007)
33. Yang, J., Gupta, R.: Energy Efficient Frequent Value Data Cache Design. In: Proceedings of MICRO, pp. 197–207 (2002)
34. Yang, J., Gupta, R., Zhang, C.: Frequent Value Encoding for Low Power Data Buses. ACM Transactions on Design Automation of Electronic Systems 9(3), 354–384 (2004)
35. Zhang, M., Asanovic, K.: Highly-Associative Caches for Low-Power Processors. In: Kool Chips Workshop, MICRO-33 (2000)
36. Zhang, Y., Yang, J., Gupta, R.: Frequent Value Locality and Value-Centric Data Cache Design. In: Proceedings of ASPLOS, pp. 150–159 (2000)

Chapter 7
Latency-Constrained, Power-Optimized NoC Design for a 4G SoC: A Case Study

Rudy Beraha, Isask'har Walter, Israel Cidon, and Avinoam Kolodny

Abstract In this chapter, we examine the design process of a network on-chip (NoC) for a high-end commercial system on-chip (SoC) application. We present several design choices and focus on the power optimization of the NoC while achieving the required performance. Our design steps include module mapping and allocation of customized capacities to links. Unlike previous studies, in which point-to-point, per-flow timing constraints were used, we demonstrate the importance of using the application end-to-end traversal latency requirements during the optimization process. In order to evaluate the different alternatives, we report the synthesis results of a design that meets the actual throughput and timing requirements of the commercial SoC. According to our findings, the proposed technique offers up to 40% savings in the total router area, 49% savings in the inter-router wiring area, and a 16% reduction of total power for our target router architecture.

7.1 Introduction

Application-specific systems on-chip (SoC) make extensive use of busses as the interconnect infrastructure. These busses are typically enhanced along product generations to match the increasing needs of the application. Such enhancements include increasing the bus frequency and width as well as enriching the bus semantics and transfer modes. By avoiding fundamental changes, the SoC architects can leverage their past experience in designing shared busses and successfully overcome the growing complexity of the design. However, in recent years, research has shown that network on-chip (NoC) is likely to replace busses in future SoCs, due to the superior performance, power, and area tradeoffs it offers as the number of modules increases [3, 5, 8, 9]. This is mainly attributed to the spatial parallelism of networks, to their short, unidirectional point-to-point wires and to their scalable architecture [4]. NoCs

I. Walter (✉)
Electrical Engineering Department, Technion - Israel Institute of Technology, Haifa 32000, Israel
e-mail: zigi@tx.technion.ac.il

C. Silvano et al. (eds.), *Low Power Networks-on-Chip*,
DOI 10.1007/978-1-4419-6911-8_7, © Springer Science+Business Media, LLC 2011

are being adopted by companies as a means to improve design productivity. As the number of modules connected to a bus increases, the physical implementation of the bus becomes very complex, and achieving the desired throughput and latency requires time-consuming custom modifications. Conversely, NoCs are designed separately from the functional units of the system to handle all foreseen inter-module communication needs. Their inherent scalable architecture facilitates the integration of the system and shortens the time-to-market of complex products.

In this chapter, we discuss the design process of an NoC for a state-of-the-art SoC. Specifically, we describe our experience in designing a cost-optimized NoC for a high-performance, power-constrained 4G wireless modem application. As the design process has many degrees of freedom creating a very large design space, finding the absolute optimal solution is an extremely difficult problem. Instead, we focus on several important choices made by the system architect while selecting some well-accepted, practical solutions to other questions.

Previous work that has dealt with the design process of the NoC often attempted to minimize power consumption and/or maximize network performance. When real applications are considered, minimizing the power consumption alone (e.g., by module mapping) is impossible, as performance constraints for each given application are to be met. Similarly, maximizing performance alone is inefficient, as excessive power might be used for improving performance beyond the needs of the application. Therefore, we look for a tradeoff between the power and performance of the NoC which is characterized by a minimal power consumption that still meets the demands of the targeted application. Moreover, in many of the studies where network latency was used as a performance goal (either as a cost function or as a constraint), the average delay of all packets over all communicating pairs was typically considered. However, in a practical SoC, different streams of communication may require different delays and therefore the overall average latency is an inappropriate measure. Consequently, the individual per-flow, point-to-point (source–destination) latencies should be accounted for to get better results.

In this chapter, we go further to suggest an improved approach: given the application that is to be used in the SoC, we utilize its functional timing requirements, which are defined by the application latency constraints. Each of those end-to-end traversal delay requirements is composed of the cumulative requirement of a sequence (or a "chain") of flows. For example, the application may require that a block of data that is generated by module A is sent to module B in order to be processed. Then, the processed data is to be sent by module B to module C for some additional processing, forming a pipeline of modules. By observing that the performance of the application is subject to the total time it would take the data to get from module A to module C, we can use this delay as the targeted performance measure, rather than specifying two separate latency constraints (for the flow from module A to module B and from module B to module C). Since pair-wise delays may be traded, the timing constraints are relaxed and the optimization tool has more freedom in its operation.

This approach is similar to re-timing of logic paths used in traditional logic synthesis tools which may "borrow time" from one pipeline stage to another to balance the timing paths and achieve high frequency of operation, as long as the total latency is not violated. Instead of moving hardware from one unit to another, the proposed technique modifies the regular NoC design flow to generate a more efficient implementation. As the main data path in SoCs is typically composed of such processing pipes, the proposed scheme is not limited to any particalur application.

The design process itself is composed of several steps: first, using simulated annealing optimization, we search for a module's placement consuming minimal power, taking into account application latency and throughput constraints. Then uniform link capacities among routers are defined to meet these performance constraints. Finally, the resulting uniform NoC is tuned by reducing the capacity of selected links.

The rest of this chapter is organized as follows: in Sect. 7.2, related work is discussed. In Sect. 7.3, we describe the characteristics of the application of the designed SoC. In Sect. 7.4, we discuss the design and optimization process of the NoC. Finally, in Sect. 7.5, we present the architecture of the NoC components (router and rate matching units) and report synthesis results. In Sect. 7.6, we summarize the chapter.

7.2 Related Work

NoC design was the subject of many papers in recent years. In particular, the problem of mapping the communicating cores onto the die has received considerable attention due to its power and performance implications. In [13], the authors propose a branch-and-bound mapping algorithm to minimize the communication energy in the system, but the resulting communication delay is not considered. In [16], a heuristic algorithm is used to minimize the average delay experienced by packets traversing the network. By allowing the splitting of traffic, an efficient implementation is found. In [14, 15], the authors use the message dependencies of the application in addition to its bandwidth requirements to find a mapping that reduces the power consumption and the application execution time. The authors of [1] use a multi-objective genetic algorithm to explore the mapping space so that a good tradeoff between power consumption and application execution time is found. While these papers use unique mapping schemes, they all use packet delay or application execution time as a quality measure rather than as an input to the mapping phase. Moreover, the metrics used does not consider the individual requirements of each pair of communicating cores, only reflecting the overall average delay or performance.

The earliest published work to consider energy efficient mapping of a bandwidth and latency-constrained NoC is [17], in which the authors specify an automated design process providing quality-of-service guarantees. Another mapping scheme that uses delay constraints as an input is described in [22]. There, a low-complexity heuristic algorithm is used to map the cores onto the chip and then routing is determined so that all constraints are met. Similarly, the mapping schemes used in [7, 11, 19] all use the per-flow, source–destination latency requirements of the application as input to the design process and find a cost-effective mapping of the cores onto the chip, satisfying the timing demands.

In this chapter, we motivate a third approach: rather than optimizing the NoC for power only and evaluating the resulting delays, or using the per-flow delay requirements as constraints during the mapping process, we use the application-level requirements which dictate end-to-end processing latencies. Wherever applicable, we replace "a chain" of point-to-point delay constraints with a single, unified constraint, describing the overall latency requirement of the application, measured from the time the first module in the chain generates the data until the last module receives the data, as explained above. In [24], the benefit that lies in this approach in terms of dynamic power is evaluated for synthetic cases. In contrast, here we use a real application and analyze the combined effect of mapping schemes and link capacity tuning alternatives on the resources occupied by the NoC. This chapter extends the work described in [2].

A major advantage of the proposed technique is that no detailed description of packet dependencies or the tasks run by each of the system cores are used in the optimization process. Instead, the proposed technique uses a standard, high-level abstract modeling of the communication within the system. This is the same approach normally used by system architects in the design processes of busses and NoCs, making this approach applicable for very complex systems. A good example of the benefit that lies in leveraging the application end-to-end temporal requirements for NoC design is given in [12], where the application data flow is analyzed to facilitate the sizing of the NoC buffers.

7.3   The Target Application

The design chosen for evaluating the conversion into the NoC architecture is a 34-module ASIC that supports all major 2G, 3G, and 4G wireless standards for use in base stations and femto cells (cell site modem – CSM), depicted in Fig. 7.1. The CSM is designed to support any of the CDMA or UMTS standards, because different markets around the world are at different points in their adoption of wireless standards.

This CSM is comprised of several subsystems that fall into three basic categories:

1. *Generic element*. These are the processor and DSP modules on chip. They are programmable and can be used for a variety of different functions.
2. *Dedicated hardware*. These blocks are designed to optimize the operations/milliwatt metric. They perform a single or a small set of operations extremely efficiently and off-load the work from the generic elements (which typically could perform the same operation but with a significant power penalty).
3. *Memory/IO*. As with most SoCs, there are memory elements and I/O modules used for information storage and communication with the outside. For the purposes of this chapter, these elements are grouped together.

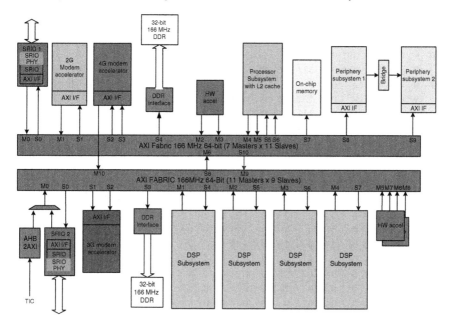

Fig. 7.1 Bus-based system architecture

In the bus-based implementation, the SoC uses a 64-bit wide, 166-MHz AXI bus at the top level. Due to design considerations such as place & route and timing closure, the interconnect fabric is segmented into two separate busses and a bridge, with approximately half the nodes on each bus. The CSM chosen for this study supports multiple modes of operation, each identified by its own bandwidth and latency requirements. In particular, it can operate in a 2G mode, 3G mode, 4G mode, and in a combination of modes for simultaneous voice and data transmissions. To find a low-cost 2D-mesh topology for the NoC, an artificial set of bandwidth requirements is generated: for each pair of nodes, the maximum bandwidth requirement it has in any of the modes of operations is selected. Similarly, we combined all the latency requirements in one table. This scenario represents the worst-case requirements in any of the modes ("synthetic worst-case" [18], "design envelope" [6]). Designing the NoC according to this scenario is likely to make it easier to meet requirements of all modes of operation in the following phases of the design, while other approaches are left for future work.

If we expect to design an NoC to replace the top-level AXI busses, it must be flexible enough to meet the bandwidth and latency requirements for each mode. However, we also do not want to overdesign the network because this will lead to area and power wastage. For the purpose of this paper, the 170 point-to-point (P2P) flows in the system running between the 16 masters and 18 slaves were described using two tables: Table 7.1 describes the bandwidth requirement between master and slave modules in the system while Table 7.2 describes the timing requirements of

Table 7.1 Bandwidth demands [Mb/s]

		S0	S1	S2	S3	S4	S5	S6	S7	S8	S9	S10	S11	S12	S13	S14	S15	S16	S17
M0	R	0	0	492	0	3	0	0	10	0	0	0	0	0	0	0	0	0	0
	W	0	0	492	0	52.6	0.6	0	10	0	0	0	0	0	37.5	0	0	0	0
M1	R	0	0	0	0	0	0	0	0	0	0	0	0	2	20	201	0.5	1	0.5
	W	0	0	0	0	52.6	0.6	0	0	0	0	0	0	35	2	2	1	202	1
M2	R	0.25	2	0	0.38	125	0	0.2	250	4	1	0	0.38	0	0	0	0	0	0
	W	0.4	2	0	5	0.1	0	0.2	125	4	1	0	0	0	0	0	0	0	0
M3	R	0	2	0	1.54	0.1	0	1	5.63	0	0	0	0	0	0	0	0	0	0
	W	0	2	0	60.7	0.1	0	1	4.38	0	0	0	40.6	0	1.88	0	0	0	0
M4	R	0	0	0	0	0	0	0	1.25	0	1	0.5	1	0	10	0	10	10	10
	W	0	0.25	0	0.25	0	2.5	0	1.25	0	1	2	1	0	10	0	10	10	10
M5	R	0	0	0	0	0	0	0	1.25	0	1	0.2	1	0	10	10	0	10	10
	W	0	0.25	0	0.25	0	2.5	0	1.25	0	1	1	1	0	10	10	0	10	10
M6	R	0	0	0	0	0	0	0	1.25	0	1	0.5	1	0	6	10	10	0	10
	W	0	0.25	0	0.25	0	2.5	0	1.25	0	1	2	1	0	10	10	10	0	10
M7	R	0	0	0	0	0	0	0	0	0	1	0.2	1	0	6	10	10	10	0
	W	0	0.25	0	0.25	0	2.5	0	0	0	1	1	1	0	10	10	10	10	0
M8	R	13	0	0	0	29.2	0	0	0	0	0	0	0	0	0	0	0	0	0
	W	11	0	0	0	0	0	0	0	0	0	0	0	0	0	0	0	0	0
M9	R	0	0	0	0	3	0	0	10	0	0	0	0	0	0	0	0	0	0
	W	29.2	0	0	0	1	0	0	10	0	0	0	0	0	0	0	0	0	0
M10	R	0	0	0	0	0	0	0	0	0	0	0	415	0	1	0	0	0	0
	W	0	0	0	0	0	0	0	0	0	0	215	400	0	1	0	0	0	0
M11	R	0	0	0	0	0	0	0	0	0	0	0	0	0	1	400	400	0	0
	W	0	0	0	0	0	0	0	0	0	0	0	0	0	1	400	400	0	0
M12	R	0	0	0	0	0	0	0	0	0	0	400	0	0	1	3.13	38.9	3.61	0
	W	0	0	0	0	0	0	0	0	0	0	400	0	0	1	0	0	200	200
M13	R	0	0	0	0	0	0	0	0	0	0	0	0	0	1	0	0	400	400
	W	9.38	0	0	36.2	0	0	0	0	0	0	0	0	0	1	0	0	400	400
M14	R	0	0	0	0	15	0	0	0.1	0	0	0	0	0	0	0	0	0	0
	W	0	0	0	0	23	0	0	0.1	0	0	0	0	0	0	0	0	0	0
M15	R	0	0	0	0	0	0	0	0	0	0	0	0	0	153	160	0	1.2	0
	W	10	0	0	0	0	0	0	0	0	0	0	0	0	0	0	0	53.8	0

R read operations, *W* write operations

point-to-point communication in the system. A third table specifies the application's end-to-end traversal delay requirements, derived from the application characteristics (Table 7.3). These tables reveal that there is a wide variability in the requirements, at both the bandwidths and the delays. For example, Master0 sends Slave2 492 Mb per second, with a latency constraint of 5,000 ns, while Master4 sends Slave16 only 10 Mb per second, but with a much tighter delay requirement of 200 ns. This variability, which is very common in modern SoCs, makes the problem of designing an efficient NoC more challenging.

Table 7.2 Point-to-point timing requirements [ns]

		S0	S1	S2	S3	S4	S5	S6	S7	S8	S9	S10	S11	S12	S13	S14	S15	S16	S17
M0	R	0	0	5,000	0	0	0	0	300	0	0	0	0	0	0	0	0	0	0
	W	0	0	5,000	0	0	0	0	300	0	0	0	0	0	0	0	0	0	0
M1	R	0	0	0	0	0	0	0	300	0	0	0	0	0	0	0	0	0	0
	W	0	0	0	0	0	0	0	300	0	0	0	0	0	0	0	0	0	0
M2	R	0	0	0	500	300	0	0	150	0	0	0	0	0	0	0	0	0	0
	W	0	0	0	500	300	0	0	150	0	0	0	0	0	0	0	0	0	0
M3	R	0	0	0	0	300	0	0	150	0	0	0	0	0	0	0	0	0	0
	W	0	0	0	0	300	0	0	150	0	0	0	0	0	0	0	0	0	0
M4	R	0	0	0	0	0	0	0	0	0	0	0	0	0	300	0	200	200	200
	W	0	0	0	0	0	0	0	0	0	0	0	0	0	300	0	200	200	200
M5	R	0	0	0	0	0	0	0	0	0	0	0	0	0	300	200	0	200	200
	W	0	0	0	0	0	0	0	0	0	0	0	0	0	300	200	0	200	200
M6	R	0	0	0	0	0	0	0	0	0	0	0	0	0	300	200	200	0	200
	W	0	0	0	0	0	0	0	0	0	0	0	0	0	300	200	200	0	200
M7	R	0	0	0	0	0	0	0	0	0	0	0	0	0	300	200	200	200	0
	W	0	0	0	0	0	0	0	0	0	0	0	0	0	300	200	200	200	0
M8	R	0	0	0	0	0	0	0	0	0	0	0	0	0	0	0	0	0	0
	W	0	0	0	0	0	0	0	0	0	0	0	0	0	0	0	0	0	0
M9	R	0	0	0	0	0	0	0	0	0	0	0	0	0	0	0	0	0	0
	W	0	0	0	0	0	0	0	0	0	0	0	0	0	0	0	0	0	0
M10	R	0	0	0	0	0	0	0	0	0	0	0	150	0	0	0	0	0	0
	W	0	0	0	0	0	0	0	0	0	0	150	150	0	0	0	0	0	0
M11	R	0	0	0	0	0	0	0	0	0	0	0	0	0	0	100	100	0	0
	W	0	0	0	0	0	0	0	0	0	0	0	0	0	0	100	100	0	0
M12	R	0	0	0	0	0	0	0	0	0	0	0	150	0	0	0	0	0	0
	W	0	0	0	0	0	0	0	0	0	0	0	150	0	0	0	0	100	100
M13	R	0	0	0	0	0	0	0	0	0	0	0	0	0	0	0	0	100	100
	W	0	0	0	0	0	0	0	0	0	0	0	0	0	0	0	0	100	100
M14	R	0	0	0	0	0	0	0	0	0	0	0	0	0	0	0	0	0	0
	W	0	0	0	0	0	0	0	0	0	0	0	0	0	0	0	0	0	0
M15	R	0	0	0	0	300	0	0	0	0	0	0	0	0	300	0	0	0	0
	W	0	0	0	0	300	0	0	0	0	0	0	0	0	300	0	0	0	0

R read operations, *W* write operations

7.4 NoC Design and Optimization

The design process of the NoC is composed of four phases:

1. Mapping the communicating modules (e.g. [1, 7, 11, 13–19, 22])
2. Trimming and adjusting the network resources to meet the application requirements [10]

Table 7.3 End-to-end traversal timing requirements [ns]

#	Mod#1	Mod#2	Mod#3	Mod#4	Req.
1	M0	S7	M3	S4	770
2	S10	M10	S11		315
3	M12	S11			150
4	S14	M11	S15		215
5	S16	M13	S17		215
6	S16	M12	S17		215
7	M2	S4			310
8	M2	S7			310
9	M4	S13			310
10	M5	S13			310
11	M6	S13			310
12	M7	S13			310
13	M0	S2			5,000
14	S7	M0			300
15	S13	M15			300
16	M2	S3			510
17	M4	S15			210
18	M4	S16			210
19	M4	S17			210
20	M5	S16			210
21	M5	S17			210
22	M6	S17			210

3. Synthesizing the network
4. Placing and routing of the NoC

The initial topology chosen for the NoC is a widely used regular 2D-mesh grid
that mitigates the concern of deadlocks and also simplifies the routing algorithm.
However, the proposed design technique is applicable to other topologies too. In
order to simplify the mapping process, all modules are considered to be of the same
size during this step of the optimization, leaving it to the place & route tool to
account for the actual placement of the chip. A more complex approach is left for
future work. In order to minimize the buffering cost and allow fast delivery of data,
wormhole switching is used.

The structure of this section is as follows: in Sect. 7.4.1, we describe three mapping alternatives to optimize communication power consumption. In Sect. 7.4.2,
we discuss two link capacity optimization schemes to further reduce the cost of
the NoC.

7.4.1 Cost-Optimized Mapping

In order to find the best 2D-mesh topology, we explore three possible optimization
goals:

1. *Power-only.* In this mapping, only the bandwidth requirements of the application are considered, while meeting the timing requirements is left for the following stages of the design process.
2. *(Power + P2P)-based placement.* Here, point-to-point latency requirements are introduced as constraints in the mapping phase.
3. *(Power + E2E)-based mapping.* Instead of specifying latency requirements for each source–destination pair, the end-to-end (E2E) traversal latency constraint of the stream of information in the application is used. For example, if data are sent from node-X to node-Y and then from node-Y to node-Z, the E2E latency is measured between node-X and node-Z. The E2E constraints are extracted from the application's characteristics and may replace some of the P2P requirements, creating a more relaxed set of constraints. A point-to-point requirement that is not a part of a larger chain is considered as an ETE traversal latency constraint.

In order to find an optimal mapping for the SoC, we define a cost function to compare different mappings. The cost function is defined as:

$$\text{Cost} = \alpha \text{AREA}_{\text{router}} + \beta \sum_{l \in \text{links}} BW_l, \tag{7.1}$$

where $\text{AREA}_{\text{router}}$ is an estimate for the total resources required to implement the router logic (accounting for each individual router number of ports and the hardware needed for the capacity it provides, which change from one mapping to another) and BW_l is the bandwidth delivered over a link l. While $\text{AREA}_{\text{router}}$ models the area and static power used by the NoC resources, the second term is commonly used to capture the dynamic power consumed by the communication (e.g. [21, 23]).

In order to search for an optimal mapping, a topology optimization tool that uses a simulated annealing (SA) algorithm was developed. The tool, which is capable of evaluating different $M \times N$ configurations for the 2D-mesh, takes as input a spreadsheet listing connectivity and bandwidth requirements between nodes. In addition, it can read a spreadsheet with latency requirements which are specified in one of two ways: (1) a list of the maximum latency allowed between any two nodes on the network; (2) a list of the E2E streams and their allowed latency (including P2P requirements that could not be replaced), i.e., the nodes a particular operation must traverse and the total latency allowed for that set of flows.

The SA algorithm starts with a random mapping of all nodes on a 2D-mesh and calculates the cost (7.1) for this initial state. It then proceeds to try and swap nodes in order to find a lower cost solution. The bandwidth spreadsheet will drive the selection of a topology as this is directly included in the cost. However, for each solution that the SA algorithm generates, the tool uses the latency spreadsheet to check if the latency requirements are met. When the requirements are not met, the solution is rejected regardless of its cost. Figure 7.2 depicts a typical example of the cost reduction behavior along the run time of the SA optimization. We use the SA tool to generate mappings using the Power-only, Power + P2P, and Power + E2E schemes, resulting in three topologies to compare.

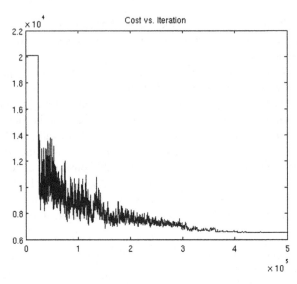

Fig. 7.2 An example of the reduction of the network cost using simulated annealing. The X-axis represents the iteration number; the Y-axis represents the cost of the network

Since run-time dynamic effects for congestion are hard to predict during the mapping phase, hop-count is often used instead (e.g., [22]). Therefore, the check reflects the length of the path traversed by the packet and the pipeline delay of the routers along that path. This approximation is accurate enough to be used by the mapping algorithm as NoCs are typically designed to operate in light loads such that congestion effects are not dominant. However, other, more elaborate analytic delay models can be equally used to account for source queuing, virtual channel (VC) multiplexing and contention [10], packetization/reassembly delay, processing time within modules, etc. Specifically, we assume a three cycle router pipeline delay, operating at 200 MHz, and account for contention in subsequent stages of the design. For the purpose of this paper, we use $\alpha = 10$, $\beta = 1$ and relative empirical weights for routers with different numbers of ports, as generated by synthesis tools. Figure 7.3 shows the mappings generated by the three schemes.

7.4.2 Setting Link Capacities

As a significant portion of the NoC area and power consumption is due to the network links, minimizing the resources used by the links has a considerable impact on the design process. In this phase, we find the required link capacity or each of the mappings generated in the mapping step.

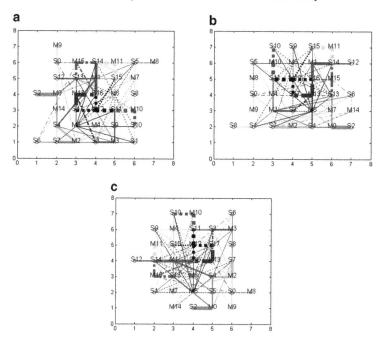

Fig. 7.3 Placement results: (**a**) Power optimized; (**b**) power optimized + P2P timing constraints; (**c**) power optimized + E2E traversal timing constraints. *Line widths* represent the relative volume of traffic

We define the total capacity of the NoC as:

$$\text{NoC}_{\text{capacity}} = \sum_{l \in \text{links}} C_l, \tag{7.2}$$

where C_l is the capacity assigned to link l, and attempt to find the minimal total capacity that would still meet all the latency constraints (same ones that were used in the mapping phase). As the mapping tool does not consider the dynamic contention within the network, this phase of the optimization process should account for all run-time effects, so that the network can deliver the required performance.

In this chapter, we consider two possible capacity allocation schemes:

1. *Uniform.* All links have the same capacity.
2. *Tuned.* An heterogeneous assignment of capacities, i.e., different links may have different capacities.

Uniform link capacity is commonly used in wormhole networks. In such cases, the process of finding the minimal capacity that meets the latency requirements using simulations is rather simple, as a single parameter (the identical capacity of all network links) is optimized. However, due to the variety of timing requirements presented by the application, this allocation causes some links to be overprovisioned.

In order to reduce the cost of the NoC, a tuned assignment can be used. In such a scheme, link capacities are individually set to much the true requirements of the system. To this end, we differentiate between two types of links: the first type is links that are used to route at least one flow which has timing requirement. The second type is those that deliver flows with no such requirements. Intuitively, it is possible to scale down links of the latter type more aggressively than those of the former type. However, it is important to note that scaling down the capacity of links that have no flows with timing requirement may hinder the delivery of flows that have latency constraints but do not traverse these links. This is due to the backpressure mechanism of wormhole switching: when a flow is slowed down in a certain router on its path, it occupies resources in other routers on its path for a longer time. Consequently, the delay of flows that share these other routers and which may have latency constraints increases. In this chapter we generate the custom, tuned allocation by scaling down the capacity found in the uniform assignment scheme: the capacity of links that are used only by flows with no timing requirements is re-assigned according to a selected utilization factor. The capacity of links that have at least one flow with latency constraint is reduced proportionally to the slack time of the flow with the lowest slack, so that reducing the capacity any further would definitely violate the timing constraint of that flow. Simulation is then used to verify that all latency constraints are met. If not, capacity is increased by a small factor and performance is verified again. In both the uniform and custom tuning schemes, links that are not used by any flow in any of the modes are completely removed.

Using an OPNET-based simulator [20] that models a detailed wormhole network (accounting for the finite router queues, backpressure mechanism, virtual channel assignment, link capacities, network contention, etc.), the basic three topologies (generated by the Power-only, Power + P2P, and Power + E2E optimizations) were simulated, using one and two virtual channels. For each case, we find the optimal network bandwidth for both the uniform and the tuned links capacity cases. Figure 7.4 illustrates the per-link capacity allocated for the mapping created by the Power-only optimization. This phase results in 12 generated networks (three basic mappings * two VC configurations * two capacity schemes). At the end of this phase, all timing requirements are met (P2P constraints in the Power-only and Power + P2P mappings, and E2E-traversal constraints in the Power + ETE generated mapping). Figure 7.5 summarizes the results, presenting the total capacity required in each of the 12 configurations.

It should be noted that tuning the capacities of links in this phase may result in arbitrary capacity values. However, the implemented hardware can support only a finite, discrete set of capacities. While setting unique link frequencies is possible, in this work, customized capacities are achieved by means of different flit sizes (32, 64, and 128 bits width). The implementation of the hardware required for translation (rate-matching blocks) is discussed in the following section. The reported results account for the cost of that hardware.

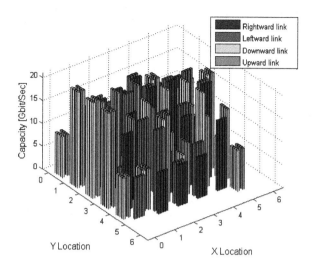

Fig. 7.4 An example of the tuning of link capacities

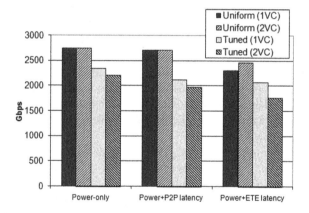

Fig. 7.5 Capacity requirements. The total capacity needed to match the requirements of the examined configuration, using one and two virtual channels

7.5 Experimental Results

In this section, we evaluate the hardware cost of each of the design choices described above. To this end, we first describe the architecute of the NoC routers. Then, we report synthesis results to compare the various alternatives.

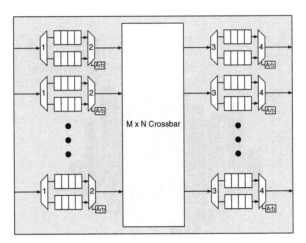

Fig. 7.6 Router architecture

7.5.1 Target Router Architecture

The basic router used for our analysis consists of an $M \times N$ switch, with M inputs and N outputs. A variable number of VCs can be implemented on each input and output. Figure 7.6 shows an implementation with two VCs per port.

Each input or output port arrow actually represents multiple signals. Packet flits are routed with two control bits, allowing flow control at each stage in the router. The VALID flag indicates that the word is valid and the WAIT flag indicates whether the receiver will accept the flit. When multiple VCs are present, there is one VALID/WAIT pair per VC.

The full data path for an Input VC consists of a demultiplexor ("1" in Fig. 7.6), a VC allocator (Alloc), some number of buffer/FIFO stages (rectangles between "1" and "2"), a multiplexor ("2"), and a VC arbiter ("Arb"). The Input VC block is shown in Fig. 7.7.

The demultiplexor ("1") diverts incoming packets to a particular input VC queue based upon an assignment made by the allocator block. The buffer stages allow separate packets to queue in parallel, thus allowing one packet to bypass another blocked packet.

The allocation block maintains a record of each incoming packet to determine if it was previously assigned a VC queue or it is a new packet that needs assignment. Each flit that comes into an input port comes from a VC in the sender. There is a VALID for each of the sender output VC. Let us call this VC-sender. When a header flit is received at an input port, the VC allocator must map the VC-sender to a VC queue that is free. To do this, the allocator maintains a list of the queues that are free. It will read an entry from the "free list" and store the mapping from VC-sender to internal VC queue. From that point on, any flits that come in on VC-sender are mapped to VC queue. When the last flit of the packet is sent out of the VC queue, the allocator will clear the mapping and make it available in the "free list."

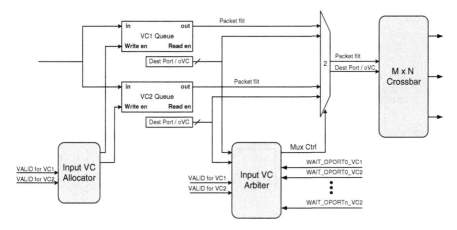

Fig. 7.7 Router input VCs

Finally, the arbiter decides which flit accesses the crossbar switch based upon the queue status at the output of the router, and controls mux "2" appropriately.

Packet flits sitting in the VC queue must be routed to the appropriate output port. This is accomplished with two levels of arbitration to decide which flits pass from input queues to output queues. The first arbitration occurs on each input port individually to determine which input VC queue passes a flit. The arbiter must tell the multiplexor ("2") which input VC queue will access the switch each clock cycle. The decision is a function of which of the input queues are not empty and whether the destination switch port (assigned output VC queue) can accept another flit (one of the WAIT_OPORTi_VCi bits). Without this first level of arbitration, the switch would grow to be $2M \times 2N$, assuming two VCs on each input and output.

The second level of arbitration is done in the crossbar switch. It arbitrates between multiple inputs accessing the same output port (despite possibly different output VCs). The mux should ensure that if a destination output queue is not ready for the next flit it will not block flits from other input VCs going to other output queues.

The data path structure of the output VCs is very similar to the input VCs. It consists of a demultiplexor ("3"), buffers (rectangles), multiplexor ("4"), and arbiter ("Arb"). Each component operates similar to the corresponding component for input VCs.

In the case of our tuned networks, a set of rate-matching blocks that will allow the transition from one packet width to another are needed. Two designs for the rate-matching blocks are shown in Fig. 7.8. The design in Fig. 7.8a converts a 128-bit packet width to a 32-bit packet width. The incoming packet is first stored in a queue. The control logic will then read the queue in groups of 32-bits and multiplex them onto the output. Because the output rate is much slower than the input rate, the control logic must also manage the WAIT signals back to the sender. This will throttle the rate of the incoming packets. We provide a mechanism to bypass the

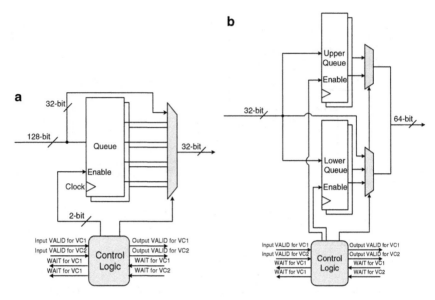

Fig. 7.8 Examples of rate-matching blocks: (**a**) 128-bit to 32-bit (**b**) 32-bit to 64-bit

queue and send 32 bits straight to the output. This is done to minimize the latency of the rate-matching block. In this case, the control logic will select the upper 32 bits of the incoming packets and mux them directly on the output. Conversely, when going from a low rate to a high rate, we use a design as shown in Fig. 7.8b. In this case we store the first valid word of the packet in the upper set of holding registers and wait for the second valid word to come in. When the second word appears, it is directed to the lower holding registers. When both upper and lower queues have data, the control logic will combine these onto one output bus and set the appropriate enable high. Here too we provide a mechanism to reduce the latency through the rate-matching block. This is done by providing the input directly to the lower mux. Similar to the 128-to-32 rate-matching blocks, the control logic must take into account the WAIT signals coming from the downstream receiver and must appropriately throttle the incoming packets.

It is worth noting that the number of queues must match the number of VCs. In our examples above, there are two sets of queues, one for each VC. The reason for this is that the rate-matching block should not block any packet from moving forward if the VC it is on is open downstream.

7.5.2 Synthesis Results

The optimization of the 1 VC network vs. the 2 VC network results in different capacity requirements for both the uniform and tuned cases. For some links, the 2 VC approach resulted in a lower link capacity because of the improved link utilization offered by the additional VC. However, the area impact of a 2 VC router must also be taken into account when choosing the best topology. Another factor to consider in the design of the network is the supported flit width. While the network bandwidth allocation algorithm allowed for any speed, the implementation of the NoC on the ASIC is limited to the clock frequencies and flit widths available in the design. For this reason, we bin the resulting router configurations into discrete categories supported on chip. We applied this binning strategy to all topologies and synthesized the network for each (the implementation is discussed in the appendix).

Figures 7.9 and 7.10 show the area results reported by the TSMC 65 nm process technology synthesis tool, separately listing the cell area and routing area. The

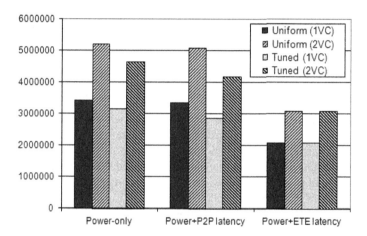

Fig. 7.9   Total router logic area. The total area consumed by routers in each of the three placement schemes

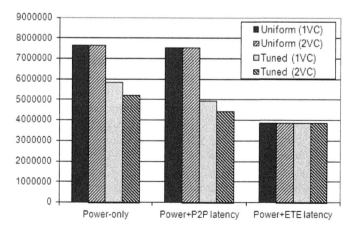

Fig. 7.10   Total wiring area. The total area consumed by inter-router wires in each of the three placement schemes

cell area includes the area taken up by the rate-matching blocks (discussed in the appendix) needed for translating one flit width to another in the network. It also accounts for the trimming of the routers, achieved by the removal of unused ports.

Analysis of the results shows that using the network capacity allocation scheme reduced the capacity of the over-provisioned links, thus saving area and power. The results also indicate that the Power + E2E latency approach provides a considerable better solution. To understand this, we must go back to mapping phase (Sect. 7.4.1). In the Power-only case, the latency requirements are completely ignored, which gives SA algorithm the most flexibility in placing the nodes on the network and finding a low-cost mapping, in terms of dynamic power consumption. When latency is included in the topology planning, the tool will reject any solution that does not meet the latency requirements. This effectively reduces the solution space for the SA algorithm, as solutions that violate the timing constraints are not allowed. Because of this, the Power + P2P scheme is the most restrictive, while in Power + E2E scheme the tool has some more flexibility in moving the nodes around as long as the latency requirement is met for the full E2E traversal path (Sect. 7.4.1).

The above explanation taken alone would imply that the Power-only case should produce the best results because the topology tool has the highest flexibility (i.e., no timing requirements have to be met). However, Figs. 7.9 and 7.10 show that the Power-only implementation has the largest area, in each VC configuration and capacity allocation scheme. To understand this, one must examine the bandwidth and latency requirements: there are some communication streams that have relatively low bandwidth but still have strict latency requirements. The nature of the topology cost function will place high bandwidth nodes close together in order to minimize the cost. When high bandwidth nodes are put close together, the other nodes get pushed further apart. As a result, some streams that require low latency will be separated by many hops. As explained above, during the link capacity tuning phase, link bandwidth is set so that all timing requirements are met. The further the distance latency-critical packets have to travel, the higher the link capacity along the path will need to be. Consequently, the Power-only case results in a very high network capacity, and thus to wider flits and an overall larger area.

In contrast, the Power-E2E scheme had the most flexibility to map nodes while at the same time making sure that the latency critical signals were relatively close together. Hence, during the network capacity allocation phase, a lower link speed could be used as compared to the Power-only case. This translates into the use of smaller flit widths for the network. Consequently, the ETE-traversal approach reduces the cell area by 25–40% (Fig. 7.9) and wiring resources by 13–49% (Fig. 7.10) compared to the traditional power + P2P mapping scheme.

Interestingly, when we consider the number of VCs, we see that one VC is preferable from an area perspective. Though using VCs reduces some of the link capacities, the savings are more than offset by the increased router sizes. Therefore, the 2 VC approach does not benefit the target application. In addition, we see that the Uniform and Tuned Power + E2E topologies have the same area. The reason for this is the binning strategy: since we are limited to 32/64/128-bit flits, our link and router selection is limited to a set of discrete choices. While it is true that the tuned

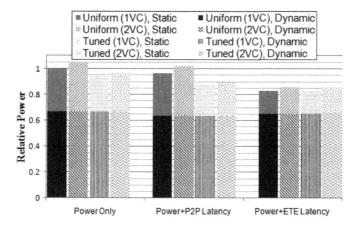

Fig. 7.11 Relative power consumption. Estimated dynamic and static power consumption in each of the three placement schemes

Power + E2E topology can run some links at a slower speed, the difference from the uniform topology is not significant enough in this case. For example, the tuned topology can reduce the speed of some links down from 15 to 14 Gbps, but given the supported flit widths, this does not change the size of the link or router we are able to choose.

Finally, Fig. 7.11 compares the estimated relative total power consumption. Static energy is based on the area reported by the synthesis tool, while dynamic energy is calculated according to the activity factor dictated by the module mapping in each configuration. As accurate, absolute power numbers can be extracted only after the place & route phase, all results are normalized by the consumption of the uniformly allocated 1 VC NoC generated by the power-only mapping scheme. According to the results, the various placement schemes resulted in very similar dynamic power consumption while static dissipation differs significantly. Interestingly, all three mapping schemes managed to generate efficient mappings, in the sense that none of them had to place heavily communicating modules far apart than they were placed by the other schemes, due to timing constraints of other streams. However, since the link capacity allocation phase trims down links more aggressively when timing constraints are easier to meet, the NoC generated by the (ETE + power)-mapping scheme consumes less area. Consequently, total power is reduced by 16% on average.

7.6 Summary and Conclusions

For a few years now, academic research is suggesting Network on-chip (NoC) as a means for providing efficient inter-modules communication within chips. Recently, a few companies have reported the usage of NoC in some prototypes and in commercial products. In this chapter, we describe our efforts to design a complex SoC around an NoC-based interconnect. In the first phase of the design, we explore three schemes to perform the placing of cores onto the chip: the first scheme only considers the power consumed by the transmission of packets while the second scheme uses the application source–destination latency constraints during the mapping phase. A third technique replaces the pair-wise requirements with application-level end-to-end latency constraints, allowing more freedom in the process of seeking a solution that minimizes power consumption.

Next, we trim redundant network resources (links, ports) and tune the bandwidth of links so that the requirements of the application are met. Finally, we synthesize the resulting networks to estimate their cost.

The main contribution of this chapter is the introduction of the end-to-end traversal delay constraints during the NoC mapping process. By replacing the source–destination requirements with end-to-end requirements wherever possible, we reduce the total area of the routers by 25–40%, the link wiring resources by 13–49% and total power by 16%, for our target architecture. In addition, we evaluate the potential benefit that lies in the implementation of links with individually assigned capacities. While we focus our analysis on a wireless modem application supporting a plethora of wireless standards and applications, such processing pipes are very typical in SoCs. Therefore, the techniques described in this paper can be used for designing and optimizing NoCs for other high-performance, power-constrained SoCs. Future work includes placing and routing the NoC and evaluating it against a bus-based system that delivers the same performance.

Acknowledgements This work was partially supported by the Semiconductor Research Corporation (SRC) and Intel Corp.

References

1. G. Ascia, V. Catania, and M. Palesi: Multi-Objective Mapping for Mesh-Based NoC Architectures, Proc. International conference on hardware/software co-design and system synthesis (CODES ISSS), 2004, pp. 182–187
2. R. Beraha, I. Walter, I. Cidon, A. Kolodny: The Design of a Latency Constrained, Power Optimized NoC for a 4G SoC, Proc. 3rd International Symposium on Networks-on-Chip (NoCs), 2009, p. 86
3. D. Bertozzi and L. Benini: Xpipes: A Network-on-Chip Architecture for Gigascale Systems-on-Chip, IEEE Circuits and Systems Magazine, 4(2), 2004, 18–31
4. E. Bolotin, I. Cidon, R. Ginosar, and A. Kolodny: Cost Considerations in Network on Chip, Integration – The VLSI Journal, 38, 2004, 19–42
5. E. Bolotin, I. Cidon, R. Ginosar, and A. Kolodny: QNoC: QoS Architecture and Design Process for Network on Chip, Journal of Systems Architecture, 50, 2004, 105–128
6. R. Gindin, I. Cidon and I. Keidar: NoC-Based FPGA: Architecture and Routing, First International Symposium on Networks-on-Chip (NoCs), 2007, pp. 253–264
7. K. Goossens, J. Dielissen, O.P. Gangwal, S.G. Pestana, A. Radulescu, and E. Rijpkema: A Design Flow for Application-Specific Networks on Chip with Guaranteed Performance to Accelerate SoC Design and Verification, Proc. Design, Automation and Test in Europe Conference (DATE), 2005, pp. 1182–1187

8. K. Goossens, J. Dielissen, and A. Radulescu: AEthereal Network on Chip: Concepts, Architectures, and Implementations, IEEE Design and Test of Computers, 2005, 414–421
9. P. Guerrier and A. Greiner: A Generic Architecture for On-Chip Packet-Switched Interconnections, Proc. Design, Automation and Test in Europe (DATE) 2000, pp. 250–256
10. Z. Guz, I. Walter, E. Bolotin, I. Cidon, R. Ginosar, and A. Kolodny: Network Delays and Link Capacities in Application-Specific Wormhole NoCs, VLSI Design, 2007, Article ID 90941, 2007, 15
11. A. Hansson, K. Goossens, and A. Radulescu: A Unified Approach to Constrained Mapping and Routing on Network-on-Chip Architectures, Proc. International conference on Hardware/ software co-design and system synthesis (CODES ISSS), 2005, pp. 75–80
12. A. Hansson, M. Wiggers, A. Moonen, K. Goossens, and M. Bekooij: Enabling Application-Level Performance Guarantees in Network-Based Systems on Chip by Applying Dataflow Analysis, Proc. IET Computers and Digital Techniques, 2009
13. J. Hu and R. Marculescu: Energy-Aware Mapping for Tile-Based NoC Architectures Under Performance Constraints, Proc. Asia South Pacific design automation (ASP-DAC) 2003, pp. 233–239
14. C. Marcon, N. Calazans, F. Moraes, A. Susin, I. Reis, and F. Hessel: Exploring NoC Mapping Strategies: an Energy and Timing Aware Technique, Proc. Design, Automation and Test in Europe Conference (DATE), 2005, pp. 502–507
15. C. Marcon, A. Borin, A. Susin, L. Carro, and F. Wagner: Time and Energy Efficient Mapping of Embedded Applications onto NoCs, Proc. Asia South Pacific design automation, 2005, pp. 33–38
16. S. Murali and G. De Micheli: Bandwidth-Constrained Mapping of Cores onto NoC Architectures, Proc. Design, Automation and Test in Europe Conference (DATE), 2004, pp. 896–901
17. S. Murali, L. Benini, and G. De Micheli: Mapping and Physical Planning of Networks-on-Chip Architectures with Quality-of-Service Guarantees, Proc. Asia South Pacific design automation (ASP-DAC), 2005, pp. 27–32
18. S. Murali, M. Coenen, A. Radulescu, K. Goossens, and G. De Micheli: A Methodology for Mapping Multiple use-cases onto Networks on Chips, Proc. Design, Automation and Test in Europe Conference (DATE) 2006, pp. 118–123
19. S. Murali, M. Coenen, A. Radulescu, K. Goossens, and G. De Micheli: Mapping and Configuration Methods for Multi-Use-Case Networks on Chips, Proc. Asia South Pacific design automation, 2006, pp. 146–151
20. OPNET modeler (www.opnet.com)
21. D. Shin and J. Kim: Communication Power Optimization for Network-on-Chip Architectures, Journal of Low Power Electronics, 2, 2006, 165–176
22. K. Srinivasan, and K.S. Chatha: A Technique for Low Energy Mapping and Routing in Network-on-Chip Architectures, Proc. Low Power Electronics and Design 2005, pp. 387–392
23. R. Tornero, J.M Orduna, M. Palesi, and J. Duato: A Communication-Aware Topological Mapping Technique for NoCs, Proc. the 14th International Euro-Par Conference on Parallel Processing, 2008
24. I. Walter, I. Cidon, A. Kolodny, and D. Sigalov: The Era of Many-Modules SoC: Revisiting the NoC Mapping Problem, Proc. 2nd International Workshop on Network on Chip Architectures (NoCArc), 2009, pp. 43–48

Part III
Future and Emerging Technologies

Chapter 8
Design and Analysis of NoCs for Low-Power 2D and 3D SoCs

Ciprian Seiculescu, Srinivasan Murali, Luca Benini, and Giovanni De Micheli

Abstract Networks-on-Chip (NoC), being a system-level interconnect, can play a major role in achieving low-power SoC designs. In many designs, the cores are grouped in to Voltage Islands (VIs). To reduce the leakage power consumption, an island containing cores that are not used in an application can be shutdown, while the other islands can still be operational. When one or more of the islands are shutdown, the interconnect should allow the communication between islands that are operational. For this, the NoCs has to be designed efficiently to allow shutdown of VIs, thereby reducing the leakage power consumption. In this chapter, we present methods to design NoC topologies that provide such a support for both 2D and 3D ICs. We show how the concept of VIs need to be considered during topology synthesis phase itself. We also make studies to show the benefits of migrating to 3D-stacked chips for realistic applications that have multiple VIs.

8.1 Introduction

Today, portable digital devices are widely available, targeting different application domains. Many of the applications demand high throughput and performance under tight power budgets, as the devices are battery powered. Over the last decade, there has been a lot of focus in tackling this important problem of developing low-power solutions for digital circuits.

The power consumption of a CMOS circuit consists of three different components [8]: (1) dynamic, (2) short-circuit, and (3) static or leakage power consumption. The dynamic or active power consumption is due to the switching of transistors. The short-circuit power consumption is due to the short-circuit current that arises when both the NMOS and PMOS transistors are active, conducting current from supply to ground. The static or leakage power consumption is present even

C. Seiculescu (✉)
LSI, EPFL, Lausanne, Switzerland
e-mail: ciprian.seiculescu@epfl.ch

C. Silvano et al. (eds.), *Low Power Networks-on-Chip*,
DOI 10.1007/978-1-4419-6911-8_8, © Springer Science+Business Media, LLC 2011

when a circuit is not switching. Several factors contribute to the leakage power consumption: sub threshold leakage, gate-induced drain leakage, gate direct tunneling leakage, and reverse-biased junction leakage [12].

With reducing transistor sizes, the leakage power consumption is becoming a significant fraction of the overall circuit power consumption. Moreover, the leakage power consumption also increases tremendously when the operating temperature of the chip increases. In fact, leakage power can be responsible for up to 40% of the total system power [12].

The devices run a variety of applications, and the utilization of the processor, memory, and hardware cores varies across the different applications. For example, a System-on-Chip (SoC) used in a mobile platform supports several applications (or use-cases), such as video display, browsing, and mp3 streaming.To reduce the leakage power consumption, cores that are not used in an application can be shutdown, while the other cores can still be operational. Power gating of designs has been widely applied in many SoCs [20]. If separate voltage lines are routed to every core, then all the cores that are not used for an application can be shutdown. However, this would require separate VDD and ground lines for each core, thereby increasing the routing overhead.

To reduce the overhead, cores are usually grouped into VIs, with cores in an island using the same VDD and ground lines [12, 20, 22, 25, 36, 41]. In [20], the importance of partitioning cores in voltage islands for power reduction is explained in detail. Several methods have been presented to achieve shutdown of islands [12, 20, 22, 25, 36, 41]. A popular technique to shutdown cores is to apply power gating using sleep transistors [12]. In such a method, sleep transistors are inserted between the actual ground lines and the circuit ground (also called the virtual ground) [12], which are turned off in the sleep mode to cut-off the leakage path. When all the cores in an island are unused for an application, the entire island can be shutdown. Many fabrication technologies provide support for VI shutdown. For example, the IBM fabrication processes CU-08, CU-65HP, and CU-45HP all support the partitioning of chips into multiple VIs and power gating of the VIs [19].

NoCs are a scalable solution to connect the cores inside chips [5, 10, 14]. NoCs consist of switches and links and use circuit or packet switching principles to transfer data across the cores. NoCs can play a major role in reducing the power consumption of the entire SoC. The design of the NoC plays an important role in allowing a seamless shutdown of the islands. When one or more of the islands are shutdown, the interconnect should work seamlessly to connect the islands that are operational. For this, the NoCs have to be designed efficiently to allow shutdown of VIs, thereby reducing the leakage power consumption.

The shutdown support presents two distinct challenges for the NoC: the NoC components should be able to handle the multiple frequencies and voltages, and the topology should allow the packets to be routed even when some islands are shutdown. Many NoC architectures support the Globally Asynchronous Locally Synchronous (GALS) paradigm and handle multiple frequencies and voltages. In [7], an architecture for a GALS NoC is presented. In [26], the authors present a physical implementation of multi-synchronous NoC. Architectures for designing

NoCs using GALS paradigm is presented in [3], and architectures for designing NoCs for GALS and DVFS operation are shown in [4]. In [32], the authors present a design methodology for partitioning an NoC into multiple islands and assigning the voltage levels for the islands.

Designing an NoC topology that meets all the application constraints has been addressed by many works. Researchers have targeted designing simple bus-based architectures [21, 33, 35] to regular NoC topologies [17, 27, 28]. Recently, several works have also addressed the issue of designing application-specific custom topologies that are highly optimized for low-power consumption and latency [1, 15, 16, 30, 34, 39, 43, 45]. However, designing a NoC topology that can support partial shutdown of the system has received less attention. In [11], the authors present approaches to route packets even when parts of the NoC have failed. A similar approach can be used for handling NoC components that have been shutdown. However, such methods do not guarantee the availability of paths when elements are shutdown. Moreover, mechanisms for re-routing and re-transmission can have a large area-power overhead on the NoC [29] and are difficult to design and verify.

One simple solution that will allow the interconnect to support shutdown of the islands would be to place the entire NoC in a separate VI. However, this is impractical, as the NoC switches could be spread across the floorplan of the chip, thereby physically spreading over multiple VIs. In this case, it is difficult to route the same VDD and ground lines across all the NoC components to keep them in a separate VI. On the other hand, if all the NoC switches are physically placed together in a single (separate) VI, then the core to switch links will become long. This would lead to large wire delay and power consumption. Moreover, this defeats the purpose of using NoC as a scalable interconnect medium.

Designing a NoC topology that not only meets application communication requirements but also allows for shutdown of islands is a challenging task. In this chapter, we present a method to design NoC topologies that allow shutdown of VIs, thereby playing a vital role in achieving a low-power design. We show how the concept of VIs need to be considered during topology synthesis phase itself.

Recently, 3D stacking of silicon layers has emerged as a promising direction for scaling [2, 6, 9, 13, 18, 23, 44]. In 3D stacking, a design is split into multiple silicon layers that are stacked on top of each other. The 3D-stacking technology has several major advantages including smaller footprint on each layer, shorter global wires, and ease of integration of diverse technologies, as each could be designed as a separate layer. A detailed study of the properties and advantages of 3D interconnects is presented in [2] and [42].

In this chapter, we also show how the NoC can be designed for 3D ICs that support VIs as well. We consolidate our works on designing NoCs for supporting VIs, presented in [37], with our works on topology synthesis for 3D ICs, presented in [31, 38]. We study how much performance and power consumption benefit can be achieved for a variety of SoC benchmarks when migrating from a 2D design to a 3D-stacked design. The rest of the chapter is organized as follows: in Sect. 8.2, we present the assumed architecture for VI shutdown and 3D integration; in Sect. 8.4, we formulate the synthesis problem for application-specific

NoC topologies supporting VI shutdown. The algorithm for synthetizing custom NoCs for 2D-Ics with VI is presented in Sect. 8.5 and the extension of the algorithm from 3D is described in Sect. 8.6. Experimental results and analysis of the results are presented in Sect. 8.7.

8.2 Architecture with Voltage Island Support

The design is partitioned into a number of voltage islands. Since for each island a separate voltage line is needed, the use of more islands will lead to difficulty in routing the different voltage lines. On the other hand, with more islands, a finer control of the shutdown mechanism can be achieved, leading to reduction in power consumption. Thus, the number of islands is chosen by the designer by carefully evaluating the tradeoffs. Usually, the designer also iterates on the number of islands and performs an architectural exploration of the design space. For a particular island count, the assignment of cores to the islands is done based on their functionality and also on their position in the floorplan. Since an entire island will be shutdown to reduce power consumption, most of the cores in an island should be active or idle at the same time for a particular application. Also, cores that have heavy communication between them need to be clustered in the same island. This is because the inter-island communication needs to traverse a frequency and voltage crossing interface, which adds to delay and power consumption. Moreover, to keep the routing of voltage lines simple, cores in an island should be located physically close to each other in the floorplan. Thus, assigning cores to islands is a multiconstrained problem and there are several works, such as [25, 32], that address this issue in more detail. In this chapter, we will only cover the complementary problem of designing a hierarchical NoC for supporting the voltage island concept. We will show how the NoC is designed for 2D and 3D ICs and show the performance benefits of 3D integration in such a scenario.

8.2.1 2D SoC Architecture

An example architecture of the 2D system that we consider for the NoC design is presented in Fig. 8.1. The assignment of the cores to different VIs is performed before the NoC design process and is taken as an input. The cores in a VI have the same operating voltage (same power and ground lines) but could have a different operating frequency. Since the inter-island communication passes through voltage and frequency converters, to reduce power consumption and latency, cores in a VI are connected to switches in the same VI. The NoC of an island (switches and links) operates in a synchronous manner, with the components using the same frequency and voltage value. This is compliant with the Globally Asynchronous, Locally Synchronous (GALS) approach, where the NoC in each island is synchronous and the

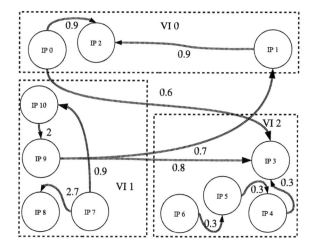

Fig. 8.1 Example input

different islands could have different frequencies and voltages. Having a locally synchronous design also eases the integration of the NoC with standard back-end placement and routing tools and industrial flows.

The cores are connected to the NoC switches by means of network interfaces (NIs) that convert the protocol of the cores to that of the network. Since the switches in a VI all operate at the same frequency, and the cores can operate at different frequencies, the NIs are also responsible for performing the clock frequency conversion. If switches from different VIs are connected together, then a frequency synchronizer has to be used on the link connecting the two switches. Bi-synchronous FIFO-based frequency converters have been proposed in literature, which can be used for the clock synchronization between switches from different VIs. The FIFOs also take care of the voltage conversion across the islands. Different VIs have different clock trees; therefore, even if they operate at the same frequency, there can be a clock skew between the islands and frequency converters still need to be used. Links connecting switches from different VIs can be routed through other VIs. Hence, for simplicity, we assume that the inter-island links are not pipelined. This restriction can be removed if the pipeline flip-flops are carefully placed in the sending or receiving switch's island.

The shutdown mechanism of an island itself could be performed in many ways. One possible approach for shutdown is the following: when an application does not require cores in a VI, the power manager decides to shutdown the island. The power manager could be either implemented in hardware or could be in the operating system. Before shutting down the island, the manager checks whether all the pending transactions on the NoC into and out of the island are completed, by interrupting or polling all the NIs. If there are no pending transactions and the cores are ready to be shutdown in an island, the power manager grounds the voltage lines to all the cores and network components in the island. In this chapter, we show a general synthesis procedure that is applicable to a system using any shutdown mechanism.

8.3   3D SoC Architecture

We assume a 3D manufacturing process based on the wafer-to-wafer bonding technology. In this, through silicon vias (TSVs) are used for establishing vertical interconnections. A vertical link requires a TSV macro on one of the layers (say the top layer), where the via cuts through the silicon wafer. In the bottom layer, the wires of the link will use a horizontal metal layer to reach the destination. For links that go through more than one layer, TSV macros are required in all the intermediate layers. However, it is important to note that the macros need not be aligned across the layers, as horizontal metal layer can be used to reach the macro at each layer as well. Stacked TSVs are not used, as the alignment of the TSVs would complicate floorplanning. The area of the TSV macros for a particular link width is taken as an input. For the synthesized topologies, our tool automatically places the TSV macros in the intermediate layers and on the corresponding switch ports. Our synthesis process automatically places the TSV macros at different layers for the different vertical interconnects.

An example of the assume architecture is presented in Fig. 8.2. From the bottom layer, the link is first routed horizontally on the metal layer and then vertically. The switch in the top layer has a TSV macro embedded for the port that is connected to this link.

8.4   Design Approach

In this section, we will first show the method for designing the NoC for 2D design to support shutdown of VIs. Then, we show the extensions for designing 3D ICs.

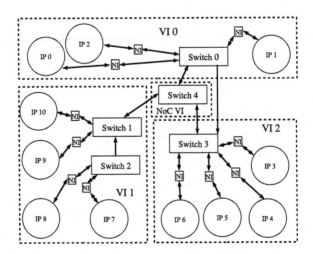

Fig. 8.2 Example architecture

8.4.1 Synthesis Problem Formulation

The synthesis procedure generates switches in each of the VIs. First, the intra-VI communication is tackled, by establishing switches in each VI and connecting the cores of the VI to the switches. Then, the inter VI traffic flows are considered and connections between switches across VIs are established. When connecting switches across VIs, we can either establish a direct connection or use switches in one or more intermediate VIs. For the latter case, we should ensure that the switches in the intermediate islands (apart from the source and destination islands) are not shutdown. A direct connection from a switch in the source island to destination island will lead to lower latency and also usually to lower power consumption. However, when there are too many inter VI flows, the size of switches may increase and the number of inter-layer links and frequency/voltage converters used may also become large. In such a case, using an intermediate VI with switches would be helpful. To ensure proper system operation, this intermediate VI should never be shutdown. The availability of power and ground lines needed to create this intermediate NoC VI is given as input and therefore the usage of the intermediate switches in the NoC VI is optional.

Routes for traffic flows that cross VI bounds are generated by the synthesis algorithm in two ways: (1) the flow can go either directly from a switch in the VI containing the source core to another switch in the VI containing the destination core, or (2) it can go through a switch which is placed in the intermediate NoC VI, if the VI is available. The switches in the intermediate VI are never shutdown. If the intermediate NoC VI is allowed, then the method will automatically explore both alternatives and choose the best one for meeting the application constraints.

The objective of the synthesis method is to determine the number of switches needed in each VI, the size of the switches, their operating frequency, and routing paths across the switches, such that application constraints are satisfied and VIs can be shutdown, if needed. The method we present here also determines whether an intermediate NoC VI needs to be used and if so, the number of switches in the intermediate island, their sizes, frequency of operation, connectivity and paths.

An example NoC design that would be an output of the design approach is presented in Fig. 8.3. As seen, the switches are distributed in the different VIs. The method produces several design points that meet the application constraints with different switch counts, with each point having different power and performance values. The designer can then choose the best design point from the trade-off curves obtained.

In order to design the NoC, we should pre-characterize the area, power, and latency models for the individual NoC components (switches, NIs, and bi-synchronous FIFOs). The models can be obtained from synthesis and place and route of the RTL code of the components using commercial tools. The generated library of the NoC components is then used during topology synthesis. As mentioned in the last section, the number of cores and their assignment to VIs are taken as inputs. Optionally, the size and position of the cores are also obtained as inputs. This floorplan information of the cores, if given, will lead to a better estimation

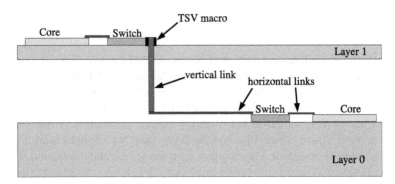

Fig. 8.3 Three-dimensional IC architecture

of the wire power consumption and delays during the synthesis process. Another input is the communication description. In the communication description, for each traffic flow, the source and destination cores are specified and also bandwidth and latency constraints are given.

Based on the inputs and models, we will show methods that synthesize different topology design points. All the topologies generated will comply with the constraints given in the input description files and may have different values for power, average latency, wire length, and switch count. A detailed description of the algorithm is given in the next section. If the size and initial positions of the cores were given as inputs, a floorplan with the NoC will also be generated. The floorplanning routine finds the best location for the NoC components and then inserts the NoC blocks as close as possible to the ideal positions, while minimally affecting the position of the cores given as input.

8.5 Synthesis Algorithm for 2D ICs with VI Shutdown

The synthesis algorithm is explained in detail in this section. From the input specifications, we construct the VI communication graph defined as follows:

Definition 8.1. A VI communication graph (VCG(V, E, isl)) is a directed graph, each vertex $v_i \in V$ represents a core in the VI denoted by isl, and the directed edge (v_i, v_j) representing the communication between the cores v_i and v_j. The bandwidth of traffic flow from cores v_i to v_j is represented by $bw_{i,j}$, and the latency constraint for the flow is represented by $lat_{i,j}$. The weight of the edge (e_i, e_j), defined by $e_{i,j}$, is set to a combination of the bandwidth and the latency constraints of the traffic flow from core v_i to v_j: $h_{i,j} = \alpha \times bw_{i,j}/max\_bw + (1 - \alpha) \times min\_lat/lat_{i,j}$, where max_bw is the maximum bandwidth value over all flows, min_lat is the tightest latency constraint over all flows, and α is a weight parameter. communicate with any other core in the same island, edges with low

---

Algorithm 1 Core-to-switch connectivity

1: Determine the frequency at which the NoC will operate in each VI and $max\_sw\_size_j$, $\forall j$
   $\in [1 \cdots N_{VI}]$
2: $min\_sw_j = |VCG(V, E, j)|/max\_sw\_size_j$, $\forall j$
3: {Vary number of switches in each VI}
4: **for** $i = 1$ to $max_{\forall j \in 1 \cdots N_{VI}} |V_j|$ **do**
5:   **for** $j = 1$ to N_{VF} **do**
6:     **if** $i + min\_sw_j < |V_j|$ **then**
7:       $k = i + min\_sw_j$
8:     **else**
9:       $k = |V_j|$
10:    **end if**
11:    Perform k min-cut partitions of $VCG(V, E, j)$.
12:   **end for**
13:   {Vary number of switches in intermediate NoC VI}
14:   **for** $k = 0$ to $max_{\forall j \in 1 \cdots N_{VI}}$ **do**
15:     Compute least cost paths for inter-switch flows using *Check_constraints* procedure.
          Choose flows in bandwidth order and find the paths.
16:     If paths found for all flows save design point
17:   **end for**
18: **end for**

---

weight (close to 0) are added between the corresponding vertices to all other vertices in the layer. This will allow the partitioning process to still consider such isolated vertices.

The value of the weight parameter α can be set experimentally or obtained as an input from the user, depending on the importance of performance and power consumption objectives.

In Algorithm 1, we present the steps required to synthesize application-specific NoCs with support for VI shutdown. The first step is to determine the frequency at which the NoC switches have to operate in each island. The minimum required frequency is determined by the highest bandwidth that has to be supported on a link from a NI to a switch. Of course, the NoC in an island can be operated at a higher frequency if the input specifications require, but a lower frequency cannot be supported as one of the cores will not be able to transfer the required bandwidth. The bandwidth available on a link is a product of the link data width and the frequency. For a fixed link width the frequency can be determined. In our synthesis procedure, without loss of generality, we fix the data width of the NoC links to a user-defined value. Please note that it could be varied in a range and more design points could be explored, which does not affect the algorithm steps.

The frequency at which the switches are operated determines the maximum size (number of inputs and outputs) that the switches can have. Since the critical path of the switch is in the crossbar, there is a direct link between the operating frequency and the size of the switch. In the algorithm, we use $max\_sw\_size_j$ to specify the maximum size that a switch can have in VI_j. Since the frequency in the different VIs is not the same, the maximum switch size will be different in different islands.

Based on the maximum size of the switches and the number of cores in a VI, the minimum number of switches that are required to generate a topology is calculated (step 2). Let N_{VI} denote the total number of VIs in the design.

To better clarify the concepts, we provide examples along with the explanation of the different steps of the algorithm.

Example 8.1. Consider the system depicted in Fig. 8.1. We will describe how the algorithm works for one design point. The design has 11 cores divided into three islands. The first step is to determine the frequency of the NoC in each VI and to calculate the maximum size of the switches in each VI. In this example, IP 7 is generating the maximum traffic in the island VI 1, with a total of 3.6 GB s^{-1} bandwidth. Let us assume that the NoC data width is set to 4 bytes. Thus, the NoC island with IP 7 should run at 900 MHz (obtained by 3.6 GB/4 B). From our NoC libraries at 65 nm, we found that a switch larger than 3×3 cannot operate at 900 MHz. Thus, we determine that the maximum switch size for this island is 3×3. As the island has four cores, we need at least two switches in the island. The minimum number of switches needed in the other islands can be calculated in a similar manner.

In steps 4–10 of the algorithm, the number of switches in each island is varied from the minimum value (computed in step 2) to the maximum number of cores in the island.

Example 8.2. Let us assume that the minimum number of switches computed in step 2 for the example in Fig. 8.3 are 1, 2, 1 for VI 0, VI 1, VI 2. We will generate design points with different switch counts, with each point having one more switches in each VI, until the number of switches is equal to the number of cores in the VI. For this example, we will explore the following points: 1,2,1, 2,3,2, 3,4,3, 3,4,4. As several combinations of switch counts in different VIs are possible, we limit to this simple heuristic.

In step 11, for the current switch count of the VI, many min-cut partitions of the VCG corresponding to the VI are obtained. Cores in a partition share the same switch. As min-cut partitioning is used, cores that communicate heavily or that have tighter latency constraints would be connected to the same switch, thereby reducing the power consumption and latency.

Example 8.3. For the design point 1,2,1, two min-cut partitions of VCG(V,E,1) are obtained. Cores IP 9 and IP 10 communicate more and belong to the same partition. Thus, they would share the same switch. Also, all flows between cores on the same switch will be routed directly through that switch.

Once the connectivity between the cores and the switches in the VIs is established, the algorithm has to find paths and open links for the inter-switch communication flows. Some of these flows also have to cross VI boundaries. For flows that have to cross VI boundaries, a link connecting a switch in the VI with the source core to a switch in the VI containing the destination core has to be found or opened. This can lead to the creation of many new links that can lead to an unacceptable increase in the switch size, violating of the $max\_sw\_size_j$ constraint. If the NoC VI is allowed, then indirect switches in the NoC VI, which are never shutdown, can be used to decrease the size of the switches in the other VIs. These switches act as indirect switches, as they are not directly connected to the cores, but only connect other switches. If the NoC VI is used, then the number of indirect switches is varied in step 14.

For each combination of direct and indirect switches, the cost of opening links is calculated and the minimum cost paths are chosen for all the flows (step 15). The traffic flows are ordered based on the bandwidth values, and the paths for each flow in the order is computed. The cost of using a link is a linear combination of the power consumption increase in opening a new link or reusing an existing link and the latency constraint of the flow. The different scenarios for setting the link costs are shown in the *Check_constraints* procedure in Algorithm 2. When opening links, we ensure that the links are either established directly across the switches in the source and destination VIs or to the switches in the intermediate NoC island. To enforce this constraint, a large cost (*INF*) is assigned to the links that are not allowed. Similarly, when the size of a switch in an island reaches the maximum value, the cost of opening a link from or to that switch is also set to *INF*. This prevents the algorithm in establishing such a link for any traffic flow. Also, when a switch is close to the maximum size (two ports less than the maximum size), a larger value than the usual cost is assigned for opening a new link, denoted by *SOFT_INF*. This is to steer the algorithm to reuse already opened links, if possible. In order to facilitate the use of the indirect switches in the intermediate NoC VI,

---

Algorithm 2 *Check_constraints*

1: {Check if the link between $switch_i$ or $switch_j$ can be used}
2: **if** $island(switch_i) = island(switch_j)$ **then**
3:    $link\_allowed$ = TRUE;
4: **else if** $island(switch_i) = src\_isl$ and $island(switch_j) = dest\_isl$ **then**
5:    $link\_allowed$ =TRUE;
6: **else if** $switch_i$ or $switch_j$ is in NoC VI **then**
7:    $link\_allowed$ =TRUE;
8: **else**
9:    $link\_allowed$ =FALSE;
10: **end if**
11: $h = island(switch_i)$ and $k = island(switch_j)$
12: **if** $size(switch_i) >= max\_sw\_size_h$ or $size(switch_j) >= max\_sw\_size_k$ or $link\_allowed$ =FALSE **then**
13:    $cost_{ij} = INF$
14: **else if** $size(switch_i) >= max\_sw\_size_h - 2$ and $switch_j$ is in NoC VI **then**
15:    $cost_{ij} = SOFT\_INF/2$
16: **else if** $size(switch_j) >= max\_sw\_size_k - 2$ and $switch_i$ is in NoC VI **then**
17:    $cost_{ij} = SOFT\_INF/2$
18: **else**
19:    $cost_{ij} = SOFT\_INF$
20: **end if**

---

the cost of opening a link between a switch that is close to the maximum size and an indirect switch is set to $SOFT\_INF/2$. Thus, when the size of a switch approaches the maximum value, more connections will be established using the switches in the intermediate NoC VI.

Example 8.4. Let us consider the switch assignment from the previous example. In this example, the highest bandwidth flow that has to be routed first is the one from IP 7 to IP 10. This will result in opening a link between Switch 2 and Switch 1. Now, let us assume that we have to find a path for a flow from IP 9 to IP 3. Because Switch 1 is close to its maximum size and we have other flows to other VIs, the algorithm will use the switch in the intermediate NoC VI. This results in opening a link from Switch 1 to 4 and another from Switch 4 to 3. The topology with the inter-switch links opened is shown in Fig. 8.3.

If for all the flows paths that do not violate the latency constraints are found, then the design point is saved. Finally, for each valid design point, the NoC components are inserted on the floorplan, and the wire lengths, wire power, and delay are calculated. The time complexity of our algorithm is $O(V^2 E^2 ln(V))$, where V is the set of cores in the design and E is the set of edges representing the communication between the cores. In practice, the algorithm runs quite fast as the input graphs typically are not fully connected.

8.6 Extension for 3D ICs

Extending the algorithm from Sect. 8.5 to generate NoC topologies for 3D-ICs is done by combining the previously presented algorithm with the algorithm for generating custom NoC topologies for 3D-ICs from [31]. In the case of the 3D algorithm from [31], the cores are assigned to layers of the 3D silicon stack. Cores can be connected only to switches in the same layer. The 3D algorithm would explore designs with different number of switches in each layer. The concept of layer is similar to the concept of VI; however, in the original 3D algorithm all layers are assumed to be synchronous.

The extension of the 3D algorithm to support shutdown of VIs as presented in Sect. 8.5 can be done under the assumption that a VI does not span across multiple layers. This assumption makes sense because it is difficult to create a synchronous clock tree that can span on multiple layers. Therefore, even if cores on different layers operate at the same frequency and voltage level, it is very likely that there will be clock skew between them and they would need to be assigned to different VIs.

The algorithm to synthesize NoCs for 3D ICs takes as input both the assignment of cores to the silicon layers of the 3D stack as well as the assignment of the cores to VIs. Also, the maximum number of links that can cross between two adjacent layers has to be given as input. This constraint is used to limit the number of TSVs that are needed to connect components on different layers in order to increase the yield.

8.7 Experimental Results

Experiments reported in this chapter are performed using the power, area, and latency models for the NoC components based on the architecture from [40]. The models are built for 65 nm technology node. We extend the library with models for the bi-synchronous voltage and frequency converters. For reference, the power consumption (with 100% switching activity), area, and maximum operating frequency for some of the components are presented in Table 8.1. The power consumption of a 32-bit link for 1-mm length was found to be $2.72\,\mu\mathrm{W\,MHz}^{-1}$. In [24], the authors show that the power consumption of tightly packed TSVs is smaller than that of horizontal interconnect by two orders of magnitude. Therefore, the impact of power consumption and delay of the vertical links is negligible as they are very short as well (15–$25\,\mu\mathrm{m}$). Under zero-load conditions, the switch delay is one cycle, an unpipelined link delay is one cycle and the voltage/frequency converter delay is four cycles (of the slowest clock).

8.7.1 Design of 2D ICs

To support VIs and shutdown of VIs, the NoC will incur an additional overhead due to the use of voltage and frequency converters and because more links need to be opened in order to support all the flows that have to cross VI boundaries. To see how much is the overhead and how it depends on the number of VIs and on the assignment of cores to VIs, we performed experiments on several benchmarks using the 2D algorithm from Sect. 8.5. The first study is performed using a realistic benchmark of a multimedia and wireless communication SoC. The benchmark has 26 cores and its communication graph is presented in Fig. 8.4 [38].

To explore the impact of core to VI assignment on the NoC overhead, we consider two ways of assigning the cores to islands. In one instance, cores that have similar functionality (e.g., shared memories that are never shutdown) or those that are meant to work together (e.g., lower level cache and the processor it services) are assigned to the same VI. The idea is to place cores that are idle at the same time in an VI. We call this assignment *logical partitioning*. This assignment is application oriented and as we will show it will incur higher communication overhead, but it also has more potential for VI shutdown. The other way to partition the cores is based on the communication description. For this assignment of cores to VIs, which we call

Table 8.1 NoC component figures

	Energy ($\mu\mathrm{W\,MHz}^{-1}$)	Area ($\mu\mathrm{m}^2$)	Freq (MHz)
Switch 4 × 4	7.2	10,000	803
Switch 5 × 5	8.4	14,000	795
Converter	0.34	1,944	1,000

Fig. 8.4 Communication
graph

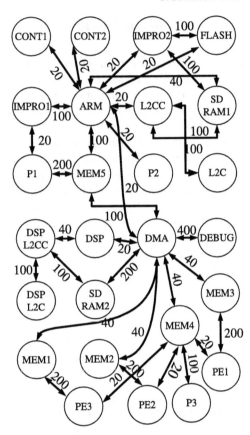

communication-based partitioning cores that have high bandwidth, communication
flows are assigned to the same VI. This assignment is communication friendly and
will reduce the overhead of the NoC, but the possibilities for VI shutdown are also
diminished.

A plot of the dynamic power consumption of the best topology for different numbers of VIs in the design is shown in Fig. 8.5a. There are two values for power
point for each VI count. One value corresponds to the case when logical partitioning is used to assign the cores to the considered number of VIs, and the other value
corresponds to the assignment by communication-based partitioning. The power
consumption values comprise the consumption on switches, links, and the synchronizers. The plot contains the two extremes: the case when there is one VI, which can
be used as reference as there is no overhead for the frequency converters. The other
extreme is for the 26 VI case when each core is assigned to its own VI. It can be
seen that for the *communication-based partitioning*, the overhead on power is not
significant until a lot of VIs are used. This is because high bandwidth flows are in
the same VI and they do not have to go through the frequency synchronizers. In the

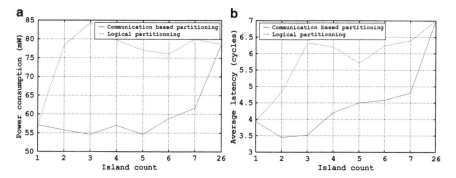

Fig. 8.5 Impact of number of VI on (**a**) power and (**b**) cycles

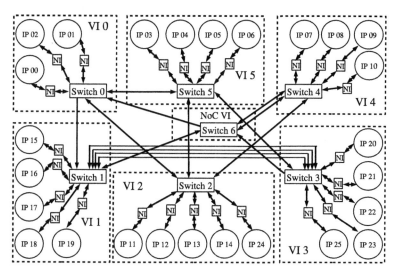

Fig. 8.6 Topology example

case of the *logical partitioning*, we have to pay some overhead in NoC dynamic power, as there are more high bandwidth flows that do go across islands.

We also consider how the average zero load latency is affected by the number of VIs for the two different assignment policies. In Fig. 8.5b, the dependence of the latency on the number of VIs is plotted. The latency value is the average zero load latency of the header flit expressed in cycles. When packets cross the islands, a four-cycle delay is incurred on the voltage-frequency converters. So with the increase in VI, the latency increases as more flows have to go through converters. However, we can see that when the assignment of cores to VIs is done based on communication the average latency does not grow so fast with the number of VIs. This is due to the assignment policy, where more flows are within the same island. A topology for the six VI *logic partitioning* case is shown in Fig. 8.6 and a floorplan example is presented in Fig. 8.7.

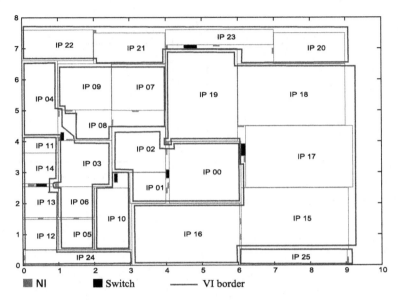

Fig. 8.7 Floorplan example

Fig. 8.8 Impact of frequency
on the switch count in the
NoC VI

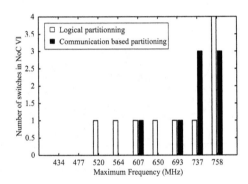

The frequency at which a switch can operate is given by the critical path inside the switch. As the bandwidth requirements of the application start to increase,
the required NoC frequencies also increase. Thus, the switches start reaching the
maximum allowed sizes to meet the frequency requirements. One way to maintain the size of switches below the threshold is the use of indirect switches in the
intermediate NoC VI. However, there is a penalty in using these switches, as a
flow going through them has to pass through one more set of voltage-frequency
converters. In Fig. 8.8, we show the number of indirect switches used for the best
power points when the frequency is varied. We see that when the bandwidth requirements are low, the intermediate island is never used. As the bandwidth requirements
start scaling, more and more indirect switches in the intermediate VI are used.

Table 8.2 Comparison on multiple benchmarks

	No VI		Multiple VIs		
	Power (mW)	Latency (cycles)	Power (mW)	Latency (cycles)	VIs
$D36\_4$	273.3	4.10	435.5	6.31	6
$D36\_6$	295.9	4.17	441.3	7.72	6
$D36\_8$	448.5	5.76	561.8	7.71	6
$D35\_bott$	112.4	5.96	117.82	6.70	6
$D65\_tvopd$	332.9	3.25	341.64	3.40	8
$D38\_tvopd$	77.43	3.31	80.12	2.62	4

The presented algorithm automatically explores the entire design space and instantiates the switches in the intermediate island when needed.

In Table 8.2, we present a comparison of power and latency between a design with no VIs and a design with multiple VIs for six benchmarks [38]. Again, we report the dynamic power consumption values of the NoC. We use different number of islands in each benchmark (average of six islands), based on the logical characteristics of the applications. The $D36\_4$, $D36\_6$, and $D36\_8$ set of benchmarks model systems with multiple shared memories on a chip. Each core communicates to 4, 6, and 8 other cores respectively with an average of 2, 3, and 4 communication flows going across the islands. In these cases, the overhead is more as there is a need for many voltage-frequency converters in order to channel all the inter-VI communication. The $D35\_bott$ benchmark has 16 processor cores, 16 private memories, and three shared memories. The processor and the private memory are assigned to the same VI. Thus, there are just the low bandwidth flows going to the shared memories that have to go across VIs. In this case, the power overhead for gating is not significant, and only latency increases since some VIs operate at a lower frequency. In case of the $D65\_pipe$ and the $D38\_tvopd$, there are 65 and 38 cores respectively, communicating in a pipeline manner and therefore there are fewer links going across VIs, resulting again in a small overhead.

For the different SoC benchmarks, we find that the topologies synthesized to support multiple VIs incur an overhead of 28% increase in the NoC dynamic power consumption. For all the benchmarks, the NoC consumes less than 10% of the total SoC dynamic power. Thus, the dynamic power overhead for supporting multiple VIs in the NoC is less than 3% of the system dynamic power. We found that the area overhead is also negligible, with less than 0.5% increase in the total SoC area. In many SoCs, the shutdown of cores can lead to large reduction in leakage power, leading to even 25% or more reduction in overall system power [12]. Thus, compared to the power savings achieved, the penalty incurred in the NoC design is negligible. Even though the packet latencies are higher when many VIs are used, the presented synthesis approach provides only those design points that meet the latency constraints of the application. Moreover, the synthesis flow allows the designer to perform trade-offs between power, latency, and the number of VIs.

8.7.2 Baseline Comparison of 2D and 3D ICs

Before we showed the effects of VIs on the NoC on 2D-ICs. Now we will show what benefits 3D integration technology can bring from the NoC perspective in designs that use VIs. But first we have to see what is the contribution that 3D technology itself brings to power savings if the circuits can be manufactured in a fully synchronous manner. For this study, we use a media benchmark, *D26_Media*. We consider a three silicon layered IC for the 3D case with the following assignment of cores to layers. The processors and DSPs with their support cores, like the caches and the hardware accelerators, are assigned to the bottom layer. The large shared memories are assigned to the middle layers and the peripherals to the top layer. We use a data width of 32 bits for the NoC links for all the experiments, matching the data width of the cores.

As previously stated, we first consider the case when both the 2D and 3D designs are implemented in a fully synchronous manner. We use this experiment as a baseline to see what is the contribution of 3D technology for power savings and as a reference for the VI overhead for the 2D and the 3D designs. As explained in Sect. 8.5, we determine the minimum required operating frequency based on the highest bandwidth requirement that any core has. For this benchmark , for a single VI, this minimum required frequency is calculated to be 270 MHz. The total NoC power consumption for the best power point for the 2D case is 38.5 mW and for the 3D design is 30.9 mW. Because the total wire length in the 3D case is significantly lower than for the 2D-IC, we obtain a 20% power savings on the NoC. This power saving is only due to the fact that wires are shorter in the 3D-IC (since the benchmark has the same number of cores), and we will show in the next section that under the restrictions of VI assignment, 3D technology can provide higher power savings when compared to 2D.

8.7.3 Comparison for Different Number of Voltage and Frequency Islands

In this section, we present a more detailed analysis and comparisons between 2D and 3D NoC designs when the cores are assigned to different VIs. We made several variations of the *D26_media* benchmark by assigning the cores to different numbers of VIs (from 1 to 7). For this comparison, we only assigned the cores to VIs using the *logical partitioning* policy as described in Sect. 8.7.1. The same VI assignment was used for both the 2D and 3D cases.

The power consumption of the best NoC topologies for different VI counts in the design for 2D-IC is presented in Fig. 8.9a. In the plot, we show the total power consumption and also the break down on the different components. A similar plot showing the power consumption of the best NoC topologies designed for the 3D-IC is presented in Fig. 8.9b. One important thing to note is that, as the number of VIs increases, the operating frequency of certain VIs can be lowered as the cores inside a VI may require less bandwidth than in another. Increasing the number of VIs increases the number of switches in a design, as there has to be at least on switch in each VI. However, when combined with the previous effect that lowers the operating frequency, we can observe that the switch power does not have a significant increase when the number of switches is increased. A similar effect can be observed on the wire power for the 2D case, as the frequency converter is placed closer to the faster switch and the link is operated at the lower frequency. In 3D, however, the switch to switch link power increases marginally with the number of islands, because of the increase in the number of switch to switch links. Power used by the frequency converters grows with the number of VIs for both cases.

In this experiment, we assume a clock skew across the different 3D layers even for a fully synchronous design, thereby leading a minimum of 3 VIs for 3D, with one for each layer. Thus, the total power consumption plotted is the same for one to three VIs in 3D. In Figs. 8.9a, b, we report the actual power values of the different topologies, and in Fig. 8.10 we show the power comparisons of topologies between the 2D and 3D cases. From the latter plot, we can see that the power savings for the case of one VI is only 10%, which is less than what is reported in the previous section. This is because, for the 3D case we actually use 3 VIs, one for each layer as the

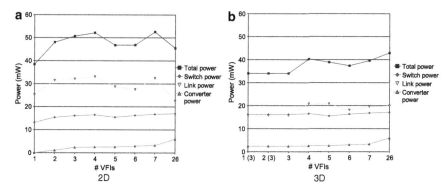

Fig. 8.9 Power for (**a**) 2D and (**b**) 3D designs

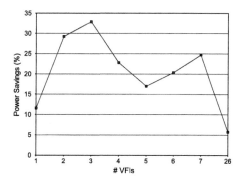

Fig. 8.10 Power savings of 3D over 2D designs

Fig. 8.11 Average zero load
latency of 2D and 3D designs

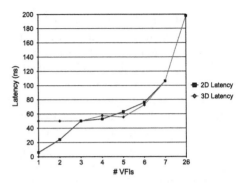

Fig. 8.12 Power savings
of 3D over 2D designs for
different benchmarks

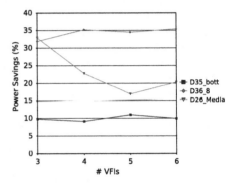

minimum number. When we compare a solution with 3 VIs in the 2D case as well, we get around 35% interconnect power savings in 3D. However, as the number of VIs increase further, the converter power starts to dominate both 2D and 3D cases. Thus, the power savings achieved by migrating to 3D reduces. Average zero load latency values for topologies designed for different number of VIs for both the 2D and 3D-ICs are shown in Fig. 8.11. Since the links are not automatically pipelined, there is not much difference between 2D and 3D designs. The average latency increases with the number of VIs because more flows have to go through frequency converters, and also because more VIs can be operated at lower frequencies.

To complete the experimental analysis, we also consider two synthetic benchmarks. The *D35_bott* and the *D36_8*, described in Sect. 8.7.1, represent two extremes. The first one has large traffic from processors to their private memories and less traffic to few shared memories. Since the private memories are close to the processors and assigned to the same VIs, the power savings in 3D are small and do not depend too much on the number of VIs, as can be seen in Fig. 8.12. The *D36_8* benchmark is the other extreme with a lot of spread traffic. In this benchmark, all cores have high bandwidth communication to eight other cores. Thus, regardless of the VI assignment, a lot of links will be required in the designs. Hence, we observe high power savings when designing the NoCs for 3D-ICs. Most realistic designs will have communication patterns between these two examples.

Fig. 8.13 Power savings
of 3D over 2D designs for
different core areas

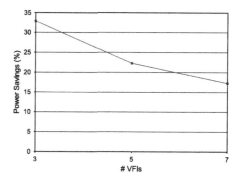

So far, we considered the case where VIs are necessary in order to be able to shutdown cores. However, if we look at future technology nodes like 32 nm and beyond, we see that the area of the region where a synchronous clock tree can be built shrinks significantly. Therefore, in large designs, VIs will be needed because it will be very expensive to build a single synchronous clock tree. To capture this effect, we perform an experiment where we increase the size of the cores in the benchmark. As we increase the size of the cores, we also have to increase the number of VIs as it would not be possible to have a single synchronous region. For example, by increasing the size of the cores by 60%, five VIs are needed instead of three if the area of a VI is kept constant. The power consumption difference between the 3D and 2D designs in percentage is shown in Fig. 8.13. As the wires are longer when the benchmark is larger, we obtain more power savings in 3D, when compared to the experiments in the previous subsection.

8.7.4 Analysis of Results

Because the 3D design has three silicon layers and frequency converters or meso-chronous synchronizers are needed when a link connects two components on different layers, migration to 3D provides little power saving (11%) with respect to a fully synchronous 2D design with no VIs. However, when more VIs are used in the design (either for functional reasons or due to technological constraints), then NoC generated for 3D SoCs consume much less power than the ones for 2D-IC. In designs with VIs, in order to support shutdown of VIs, more links are needed to route all communication flows that go across VI boundaries; hence, the shorter wire lengths in 3D designs result in considerable power savings (up to 32%). If we increase the number of VIs too much, then the wires become more segmented in 2D as well and get shorter. So, the power saving achieved in 3D NoCs starts to drop. The reduction in the power saving of 3D NoCs with the number of VIs is also due to the fact that the frequency converter power consumption becomes more significant when the number of VIs is increased. There is no significant reduction of the average zero load latency for topologies designed for 3D-ICs. This is because the average zero load delay is dominated by the four cycle delay to cross the frequency converters. The area overhead due to the insertion of TSVs in 3D is negligible, as the TSV macros occupy less than 2% area when compared to the area of the cores.

8.8 Conclusions

Leakage power consumption is becoming a large fraction of the total power consumption in ICs. In order to reduce the leakage power consumption, cores that are not used in an application can be shutdown. For ease of routing signals, cores are grouped into voltage islands and when cores in an entire island are unused, the island can be shutdown. The design of the NoC plays an important role in allowing a seamless shutdown of the islands. The NoC topology should be designed such that even when some islands are shutdown, communication across the different islands that are operational should be possible. In this chapter, we showed how the NoC topology can be designed to achieve this. We applied the methods to design NoCs for both 2D and 3D ICs. Our studies on several benchmarks show a significant reduction in NoC power consumption in migrating to a 3D technology, especially for designs with many voltage islands.

Acknowledgements We would like to acknowledge the financial contribution of CTI under project 10046.2 PFNM-NM and the ARTIST-DESIGN Network of Excellence.

References

1. T. Ahonen, D. Signza-Tortosa, H. Bin, J. Nurmi, Topology Optimization for Application Specific Networks on Chip, Proceedings of SLIP, pp. 53–60, Feb 2004
2. K. Banerjee, S. J. Souri, P. Kapur, K. C. Saraswat, 3-D ICs: A Novel Chip Design for Deep-Submicrometer Interconnect Performance and Systems-on-Chip Integration, Proceedings of the IEEE, 89(5):602, 2001
3. E. Beigne, P. Vivet, Design of On-Chip and Off-Chip Interfaces for a GALS NoC Architecture, Proceedings 12th IEEE Intl Symposium on Asynchronous Circuits and Systems (ASYNC 06), IEEE CS Press, 2006, pp. 172–181
4. E. Beigne, F. Clermidy, S. Miermont, P. Vivet, Dynamic Voltage and Frequency Scaling Architecture for Units Integration within a GALS NoC, Proceedings of the Second ACM/IEEE International Symposium on Networks-on-Chip, pp. 129–138, April 07–10, 2008
5. L. Benini, G. De Micheli, Networks on Chips: A New SoC Paradigm, IEEE Computers, pp. 70–78, Jan 2002
6. E. Beyne, The Rise of the 3rd Dimension for System Intergration, Interconnect Technology Conference, 2006 International, pp. 1–5
7. T. Bjerregaard, S. Mahadevan, R. G. Olsen, J. Sparsoe, An OCP Compliant Network Adapter for GALS-based SoC Design Using the MANGO Network-on-Chip, Proceedings 2005 International Symposium on 17-17 Nov 2005 pp. 171–174
8. A. P. Chandrakasan, S. Sheng, R.W. Brodersen, Low-Power CMOS Digital Design, IEEE Journal of Solid-State Circuits, 27(4):473–484, 1992
9. J. Cong, J. Wei, Y. Zhang, A Thermal-Driven Floorplanning Algorithm for 3D ICs, ICCAD, Nov 2004, pp. 306–313

10. G. De Micheli, L. Benini, Networks on Chips: Technology and Tools, Morgan Kaufmann, CA, First Edition, July 2006
11. T. Dumitras, S. Kerner, R. Marculescu, Towards on-chip fault-tolerant communication, ASP-DAC 2003, pp. 225–232
12. F. Fallah and M. Pedram, Standby and Active Leakage Current Control and Minimization in CMOS VLSI Circuits, IEICE Trans. on Electronics, pp. 509–519, Apr 2005
13. B. Goplen and S. Sapatnekar, Thermal Via Placement in 3D ICs, Proceedings of the International Symposium on Physical Design, p. 167, 2005
14. P. Guerrier, A. Greiner, A Generic Architecture for On-Chip Packet Switched Interconnections, Proceedings of the Conference on Design, Automation and Test in Europe, pp. 250–256, March 2000
15. A. Hansson, K. Goossens, A. Radulescu, A Unified Approach to Mapping and Routing on a Combined Guaranteed Service and Best-Effort Network-on-Chip Architectures, Technical Report No: 2005/00340, Philips Research, April 2005
16. W. H. Ho, T. M. Pinkston, A Methodology for Designing Efficient On-Chip Interconnects on Well-Behaved Communication Patterns, HPCA, 2003
17. J. Hu, R. Marculescu, Exploiting the Routing Flexibility for Energy/Performance Aware Mapping of Regular NoC Architectures, Proceedings of the Conference on Design, Automation and Test in Europe. March 2003, pp. 10688–106993
18. W.-L. Hung, G. M. Link, Y. Xie, N. Vijakrishnan, M. J. IRwin: Interconnect and Thermal-Aware Floorplanning for 3D Microprocessors, Proceedings of the ISQED, March 2006, pp. 98–104
19. IBM ASIC Solutions, http://www-03.ibm.com/technology/asic/index.html
20. D. Lackey, P. S. Zuchowski, T. R. Bednar, D. W. Stout, S. W. Gouls, J. M. Cohn, Managing power and performance for System-on-Chip designs using Voltage Islands, Proceedings of the ICCAD 2002, pp. 195–202
21. K. Lahiri, A. Raghunathan, S. Dey, Design Space Exploration for Optimizing On-Chip Communication Architectures, IEEE TCAD, 23(6):952–961, 2004
22. L. Leung, C. Tsui, Energy-Aware Synthesis of Networks-on-Chip Implemented with Voltage Islands, Proceedings of DAC 2007, pp. 128–131
23. S. K. Lim, Physical Design for 3D System on Package, IEEE Design and Test of Computers, 22(6):532–539, 2005
24. I. Loi, F. Angiolini, L. Benini, Supporting Vertical Links for 3D Networks On Chip: Toward an Automated Design and Analysis Flow, Proceedings of Nanonets 2007, pp. 23–27
25. Q. Ma, E. F. Y. Young, Voltage Island Driven Floorplanning, Proceedings of ICCAD 2007, pp. 644–649
26. I. Miro-Panades, et al., Physical Implementation of the DSPIN Network-on-Chip in the FAUST Architecture, Networks-on-Chip, 2008. Second ACM/IEEE International Symposium on 7–10 April 2008 pp. 139–148
27. S. Murali, G. De Micheli, Bandwidth Constrained Mapping of Cores on to NoC Architectures, Proceedings of the Conference on Design, Automation and Test in Europe, 2004, pp. 20896–20902
28. S. Murali, G. De Micheli, SUNMAP: A Tool for Automatic Topology Selection and Generation for NoCs, Proceedings of the DAC 2004, pp. 914–919
29. S. Murali, T. Theocharides, N. VijayKrishnan, M. J. Irwin, L. Benini, G. De Micheli, Analysis of Error Recovery Schemes for Networks-on-Chips, IEEE Design and Test of Computers, 22(5):434–442, Sep–Oct 2005
30. S. Murali, P. Meloni, F. Angiolini, D. Atienza, S. Carta, L. Benini, G. De Micheli, L. Raffo, Designing Application-Specific Networks on Chips with Floorplan Information, ICCAD 2006, pp. 355–362
31. S. Murali, C. Seiculescu, L. Benini, G. De Micheli, Synthesis of Networks on Chips for 3D Systems on Chips. ASPDAC 2009, pp. 242–247
32. U. Y. Ogras, R. Marculescu, P. Choudhary, D. Marculescu, Voltage-Frequency Island Partitioning for GALS-based Networks-on-Chip, Proceedings of DAC, June 2007

33. S. Pasricha, N. Dutt, E. Bozorgzadeh, M. Ben-Romdhane, Floorplan-aware automated synthesis of bus-based communication architectures, Proceedings of DAC, pp. 65–70 June 2005
34. A. Pinto, L. Carloni, A. Sangiovanni-Vincentelli, Constraint-Driven Communication Synthesis, Proceedings of DAC, pp. 783–788, June 2002
35. K. Ryu, V. Mooney, Automated Bus Generation for Multiprocessor SoC Design, Proceedings of the Conference on Design, Automation and Test in Europe, pp. 282–287, March 2003
36. A. Sathanur, L. Benini, A. Macii, E. Macii, M. Poncino, Multiple Power-Gating Domain (multi-VGND) Architecture for Improved Leakage Power Reduction, Proceedings of ISLPED 2008, pp. 51–56
37. C. Seiculescu, S. Murali, L. Benini, and G. De Micheli, NoC Topology Synthesis for Supporting Shutdown of Voltage Islands in SoCs. In Proceedings of the 46th Annual Design Automation Conference (DAC 2009), pp. 822–825, 2009
38. C. Seiculescu, S. Murali, L. Benini, G. De Micheli, SunFloor 3D: A Tool for Networks on Chip Topology Synthesis for 3D Systems on Chip, 2009, pp. 9–14
39. K. Srinivasan, K. S. Chatha, G. Konjevod, An Automated Technique for Topology and Route Generation of Application Specific On-Chip Interconnection Networks, Proceedings of ICCAD 2005, pp. 231–237
40. S. Stergiou, F. Angiolini, S. Carta, L. Raffo, D. Bertozzi, G. De Micheli, × pipesLite: A Synthesis Oriented Design Library for Networks on Chips, Proceedings of the Conference on Design, Automation and Test in Europe 2005, pp. 1188–1193
41. Y.-F. Tsai, D. Duarte, N. Vijaykrishnan, M.J. Irwin, Implications of Technology Scaling on Leakage Reduction Techniques, DAC 2003
42. R. Weerasekara, L.-R. Zeng, D. Pamunuwa, H. Tenhunen, Extending Systems-on-Chip to the Third Dimension: Performance, Cost and Technological Tradeoffs, Proceedings of ICCAD, 2007, pp. 212–219
43. J. Xu, W. Wolf, J. Henkel, S. Chakradhar, A Design Methodology for Application-Specific Networks On-Chip, ACM Transactions on Embedded Computing Systems (TECS), 5(2): 263–280, 2006
44. P. Zhou, Y. Ma, Z. Li, R. P. Dick, L. Shang, H. Zhou, X. Hong, Q. Zhou, 3D-STAF: Scalable Temperature and Leakage Aware Floorplanning for Three-Dimensional Integrated Circuits, ICCAD, Nov 2007, pp. 590–597
45. X. Zhu, S. Malik, A Hierarchical Modeling Framework for On-Chip Communication Architectures, ICCD 2002, pp. 663–671, Nov 2002

Chapter 9
CMOS Nanophotonics: Technology, System Implications, and a CMP Case Study

Jung Ho Ahn, Raymond G. Beausoleil, Nathan Binkert, Al Davis,
Marco Fiorentino, Norman P. Jouppi, Moray McLaren, Matteo Monchiero,
Naveen Muralimanohar, Robert Schreiber, and Dana Vantrease

Abstract The latency, bandwidth, and power consumption of on-chip interconnection networks are central concerns in the design of multi- and many-core
microprocessors. When the global network-on-chip (NoC) is electrical, the power
consumption and the limited connectivity caused by difficulties associated with
global wires will limit network performance due to power or topology constraints
unless applications can be written, which only require nearest neighbor communication. This is a highly unlikely scenario, and these performance and power barriers
will become more severe with shrinking process technology and increased core
counts. Emerging CMOS nanophotonic technologies provide a compelling alternative to traditional all-electronic NoCs. Wire power consumption is fundamentally
linear with wire length. Due to the low loss nature of waveguides, optical data
transmission primarily consumes energy at the endpoints where optical to electrical
(OE) and electrical to optical (EO) conversion takes place. Therefore, the energy
required to transport data is relatively independent of path length for path lengths of
interest for NoC-based systems. Additionally, the use of wave division multiplexing
can be exploited to improve per lane bandwidth. Signal integrity limitations make
this option intractable for electrical NoCs. The result is that nanophotonic NoCs can
provide both higher throughput and lower power consumption than all-electrical
NoCs. This chapter introduces CMOS nanophotonic technology and considers
its use in photonic chip-wide networks enabling many-core microprocessors with
greatly enhanced performance and flexibility while consuming less power than their
electrical counterparts. It provides, as a case study, a design that takes advantage
of CMOS nanophotonics to achieve ten-teraop performance in a 256-core 3D chip
stack, using optically connected main memory, very high memory bandwidth, cache
coherence across all cores, no bisection bandwidth limits on communication, and
cross-chip communication at very low latency with cache-line granularity.

J.H. Ahn (✉)
Seoul National University, Sumon, Gyeonggi-do, korea
e-mail: gajh@snu.ac.kr

C. Silvano et al. (eds.), *Low Power Networks-on-Chip*,
DOI 10.1007/978-1-4419-6911-8_9, © Springer Science+Business Media, LLC 2011

9.1 Introduction

The continuing feature size reduction of CMOS technology offers the possibility of increasing the core count of chip multiprocessors, potentially doubling every 18 months [5]. As core count grows into the hundreds, the main memory bandwidth required to support concurrent computation on all cores will increase by orders of magnitude. Unfortunately, the ITRS roadmap [42] predicts only a small increase in pin count ($< 2x$) over the next decade, and pin data rates are increasing slowly. This creates a significant main memory bandwidth bottleneck and will limit performance. Similarly, as core counts increase, a commensurate increase in on-chip bandwidth will be required to support core to core communication and core to memory controller communication. Evidence suggests that many-core systems using electrical NoCs may not be able to meet these high bandwidth demands while maintaining acceptable performance, power, and area [28].

The difficulty in creating high-bandwidth NoCs for future chip multiprocessor (CMP) systems stems from the scaling characteristics of wires [20]. As device technology shrinks, minimum-sized transistors shrink linearly. Wires also become smaller both in height and width, but the rate at which they shrink is slower. This is particularly true in the upper layers of metal, which show little or no size reduction between process steps. It is useful to characterize wires in three classes: local, intermediate, and global. Local wires typically are routed on lower layers of metal and are used to interconnect subsystems within an individual core, such as an ALU, FPU, memory controller and register file. Local wires also tend to be relatively short and their length, power, and bandwidth properties tend to scale well with process shrinks since the subsystems become commensurately smaller. Intermediate wires occupy the middle metal layers and are used to connect nonadjacent subsystems within the core. Intermediate wires can be a problem but the use of low-K dielectric materials and the length reduction due to a smaller core size combine to mitigate this problem for now. Global wires occupy the upper metal layers and are used to distribute power, ground, and clock signals. They can also be used for NoCs supporting inter-core and core to main memory communication. Global wires are definitely a power, latency, and bandwidth problem. We restrict our subsequent focus to the global wire problem.

In the absence of high off-chip bandwidth and high bisection bandwidth in the NoC, program performance is limited except for programs that exhibit very strong reference locality. Programming with this limitation has been done for some applications, but it is very difficult to do in general, and in many cases the required locality cannot be achieved. Cryptanalysis is an example of an application that demands high cross-system bandwidth. In fact, the trend in data-centric computing is to use very high level programming frameworks, such as MapReduce, that are cross-system communication intensive.

Chip-wide electrical NoC designers have two choices. They can either use global wires or not. If global wires are not used, then the interconnection topology will be limited to nearest neighbor communication which limits topology choices to a two-dimensional mesh or torus. The advantage of either topology choice is that communication can be handled by relatively short local or intermediate metal layers. The disadvantage is that non-neighbor communication will require more hops and the intermediate routers will cause increased power consumption, higher latency, and reduced cross-system bandwidth. The use of global wires will enable a richer set of topology and routing algorithm choices. However, as technology scales, global wires become increasingly expensive, as their lengths remain nearly constant, while local and intermediate wire lengths shrink.

In a nutshell, global wires do not sufficiently shrink with improved process technology, and as more metal layers are added, they may even grow slightly. Wire capacitance is linearly dependent on wire length and both power and latency vary linearly with capacitance. Bandwidth also remains relatively constant since the number of wires that can occupy a particular cross-section is effectively constant [32]. In general, latency is quadratic with wire length in unrepeated wires, but is roughly linear with length in repeated wires. However, this latency reduction comes at the cost of increased repeater power. A more detailed discussion of these issues is covered in Sect. 9.3. Keeping cross-chip delay roughly constant will cause the global interconnect to begin to dominate chip power consumption [36].

Integrated nanophotonics provides an opportunity to solve the NoC bandwidth and power problems associated with wires. A similar opportunity exists to solve the off-chip bandwidth problem imposed by external package pins and their signalling rates, which are not projected to scale with core count. This cannot be achieved where the conversion to and from the optical domain is performed with external transceivers. In order to fully benefit from photonics, it is necessary to bring light directly to the chip. Recent developments in integrated nanophotonics have demonstrated that all the components of an optical interconnect system (modulation, transmission, and detection) can be implemented in technologies which are compatible with CMOS manufacturing processes [29]. Nanophotonic NOC technology has the potential to provide substantial improvements in both bandwidth and power when compared to electrical approaches. Optical transmission media typically have very low loss compared to electronic interconnects. Unlike wire-based interconnects, the required energy per bit for optical communication is largely independent of path length since optical communication consumes energy at the path endpoints.

The telecom industry has chosen to use optics for long-haul networks. As process technology continues to shrink, so does the definition of "long." Soon, the computer industry may be able to deploy optical NoCs. Initial photonic device demonstrations have shown the potential for optical NoCs at energy levels of approximately 200 fJ per bit [45], compared with 2 pJ per bit [37] for electrical NoCs. Photonic intra-chip communication may be possible if the energy per bit can be further decreased [31].

Since information is ultimately sourced, received, and processed in the electronic domain, the interconnect path length where optics wins depends on the energy required to convert to and from the optical domain. A crossover point of as low as 50 μm has been predicted for the most aggressive optical technologies [31].

For high-speed off-chip electronic interconnects, modulation rates are close to the maximum channel bandwidth. Power hungry pre- and post-equalization is frequently required to maintain signal integrity. In contrast, the carrier frequency for optical communication is many orders of magnitude higher than the modulation frequency. The large disparity between modulation rates and carrier frequency means that optical NoCs are freed from the signal integrity considerations that limit the data rate of electronic NoCs. This has the benefit that in optical systems, modulation rates can be scaled without redesigning the transmission path. The optical NoC transmission path is also unburdened by considerations of electromagnetic interference and crosstalk.

The very high carrier bandwidth of optical systems can also be exploited through wavelength division multiplexing (WDM). In the context of NoC communication, this can allow all the lanes of a parallel data bus to be carried on a single optical fiber or waveguide, with each lane modulated on a separate wavelength of light. WDM becomes particularly attractive in the context of integrated photonics, where the incremental cost of adding additional detectors and modulators to a lane is potentially very low. WDM can also be used in switching applications where multiple nodes are connected to a common optical communication medium. The use of tunable transmitters or detectors allows nodes to select between lanes, thereby creating an optical circuit switch. As current electronic networks for computer systems are almost always packet switches rather than circuit switches, substantial architectural modification may be necessary to exploit this capability.

Intra-chip optical communication is possible using the same technologies as proposed for inter-chip communications. In this case, the transmitters and detectors are directly connected by waveguides on one or more layers of optical interconnect. Free-space optical communication has also been proposed for intra-chip communication applications [52].

In this chapter, we consider the requirements for optical interconnects for NoCs. We review the state of the art in optical light sources, modulators, detectors, and waveguides, the four principle components of an optical interconnect. We restrict our review to device technologies that have the potential for large-scale integration, and therefore their compatibility with CMOS processes, either directly through the addition of further process steps or using 3D die stacking. We then consider the implications of photonic communication for NoC-based multi- and many-core architectures. Finally, we examine a case study of a 256-core, photonically connected CMP. We develop an NoC architecture that fully exploits the capability of a particular choice of the CMOS nanophotonics technology options. We analyze the performance of the optically connected CMP compared to a reference electronic design based on an electronic 2D mesh both in terms of power and performance.

9.2 CMOS Nanophotonic Technologies

Optical interconnects have been successful for long distance data transmission and are becoming increasingly cost-effective for shorter distances. They are therefore an obvious candidate to investigate as a solution for on-chip networks. A compelling advantage of optical interconnects is the ability to increase the bandwidth of a single physical channel using dense wavelength division multiplexing (DWDM).

Given the power overheads incurred by serialization and deserialization (SER-DES), it is unlikely that on-chip interconnects will be driven at rates higher than twice the clock speed. Since active power is linear with clock frequency, clock speeds are not likely to exceed 5 GHz in the next decade [42]. With these limitations in mind, it becomes clear that electrical interconnects cannot provide the high connectivity necessary for free programmers from locality concerns in a cost- and power-effective way. Similar limitations would apply for a coarse wavelength division multiplexing (CWDM) optical interconnect. A DWDM optical approach remains, but it presents significant implementation challenges.

9.2.1 Overview

To achieve on-chip bandwidths of the order of $10\,\mathrm{TB\,s^{-1}}$, one would require a DWDM interconnect with 64 wavelengths per waveguide, a $10\,\mathrm{Gb\,s^{-1}}$ channel data rate, and approximately 250 waveguides in the network. The feasibility of such on- and off-chip photonics will require integrated photonic circuits that are compatible with conventional CMOS fabrication processes. Recent progress in silicon photonics has shown that most of the devices needed for such networks are indeed possible. In addition, when economies of scale are taken into account, it is reasonable to believe that silicon-based photonic integrated circuits can be produced at costs that are competitive with the electrical alternatives.

A complete photonic NoC requires a number of elements. Typically one needs a source of light, a modulator that can take data from the electrical domain and encode that data in the optical domain, a waveguide to transport the light across the chip, and a detector that decodes the optical data and transfers it back to the electronic domain. In the general case, a method will be necessary to connect the photonic NoC to an off-chip photonic network, and therefore some form of connector between on-chip waveguides and the off-chip transport layer will also be required. These photonic devices, with the possible exception of the optical power source, are built in a silicon layer, which may be integrated with CMOS electronics. Note that manufacturers of transceivers for telecom applications have recently started to integrate silicon photonics and electronics in the same layer [34].

The telecom approach to technology integration will probably not be particularly useful for a full-scale photonic NoC for a number of reasons. First, CMOS logic and silicon photonics require fairly different processes and it seems unlikely that both processes can be optimized at the same time. Second, in a multi-core processor, the logic circuits will occupy a large area leaving little room for the photonic components. Most likely an on-chip photonic network will require a dedicated layer. This layer can be either a dedicated chip in a 3D chip stacking scheme (as advocated by us [46], see [10] for the availability of the 3D technology) or a polycrystalline silicon layer built on the back of a logic CMOS chip [7]. Such separate-layer schemes may allow for some circuit elements, such as analog drivers for the photonic devices, to be integrated into the photonic layer.

9.2.2 Sources

There are two possible modulation schemes in an on-off keying (OOK) photonic link: direct modulation and external modulation. In a direct modulation scheme, the light source, typically a laser or an LED, is directly switched on and off. In an external modulation scheme, one uses a continuous-wave (cw) optical power source and a modulator to obtain the same result. Direct modulation is well suited for single-wavelength and CWDM systems, but we believe that for a DWDM system with multiple connections, external modulation is preferable because the narrow-band lasers can be shared between multiple photonic links.

Lasers are a key component for a silicon photonic integrated circuit (Si-PIC). Silicon light sources are very hard to build because of the material's indirect band-gap properties. Therefore, a III-V laser will need to be used to generate light either as an off-chip or as an on-chip hybrid laser.

Multi-wavelength lasers with precisely controlled frequency spacings have recently been built. One advantage of these lasers is that if only one of the frequency channels is servo-locked to an on-chip standard cavity, then all of the other frequency modes will track the controlled mode. One possible approach is the Fabry–Perot comb laser based on quantum dots [26], which has already been used to demonstrate a bit-error-rate of 10^{-13} at $10\,\mathrm{Gb\,s^{-1}}$ over ten longitudinal modes [19]. Another possible approach is the mode-locked hybrid Si/III-V evanescent laser [24], which uses a silicon-waveguide laser cavity wafer-bonded to a III-V gain medium. Any temperature change in the environment will cause approximately the same refractive index shift in the laser cavity and the silicon waveguides and resonators that form the DWDM network. This simplifies wavelength locking. A wavelength locking scheme that is robust against temperature changes is one of the key implementation challenges for a DWDM NOC.

To implement a DWDM network, it is also necessary to build frequency selective modulators and detectors. A DWDM modulator can, in-principle, be built using wavelength-independent modulators (such as Mach–Zehnder [18] or electro-optic modulators [53]) and add-drop filters. This solution is not easily implementable, given the large area of non-resonant add-drop filters such as array waveguide gratings (AWGs). An alternative solution is to use resonant elements to multiplex and demultiplex the signals. This solution is more complex to implement because resonant elements are more sensitive to environmental changes such as temperature and require very strict fabrication tolerances.

We advocate the use of silicon micro-ring resonators [49] because of their small size, high quality factor (Q), transparency to off-resonance light, and small intrinsic reflections. Using injected charge, the refractive index of the micro-ring can be

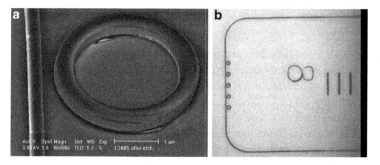

Fig. 9.1 (**a**) An SEM picture with 40°-titled view of a micro-ring resonator with a 1.5 μm radius coupled to a waveguide with an optimized (reduced) width. (**b**) A microscope picture of cascaded micro-ring resonators coupled to a U-shaped waveguide at the edge of the chip

changed to blue-shift the fundamental frequency of the cavity. This mechanism can be used to move the ring into or out of resonance with an incident light field, thus providing a mechanism for electro-optic modulation [49] in an OOK scheme. When not modulating, the rings are in the "OFF" position and negligibly interfere with light transmission. A slower tuning mechanism is provided by temperature: increasing the ring temperature red-shifts the frequency and this mechanism can be used to track slow changes in the laser frequency. Additional functionality can be obtained by using a variable-wavelength add-drop filter [35, 43] as a switch. A ring with a radius of 1.5 μm and an intrinsic quality factor Q of 18,000 has been demonstrated recently by our group [51]. This quality factor only takes into account intrinsic losses in the ring; when losses caused by the ring coupler are added, the result is a "loaded" Q of 9,000. Effective mode volumes for these rings are approximately $1.0 \, \mu m^3$ [51] . The measured quality factor is close to the maximum achievable Q, given silicon waveguide bending losses. Figure 9.1a shows a fabricated micro-ring and its associated waveguide, cascaded silicon micro-ring resonators are shown in Fig. 9.1b. These can be used as a modulator or filter bank in a nanophotonic NoC.

9.2.3 Waveguides, Splitters, Couplers and Connectors

A Si-PIC requires a number of passive components. The transport layer is built out of silicon-on-insulator (SOI) waveguides with measured losses as low as $0.1 \, dB \, cm^{-1}$ [17]. Other useful components include splitters, that can be used to distribute power and realize broadcast networks, and couplers, that are used to couple light on- and off-chip. Evanescent splitters have been studied in detail both theoretically and experimentally and are well understood. For coupling light on- and off-chip, we advocate the use of grating couplers [3,41]. Grating couplers allow one to couple light from a standard optical fiber into a SOI waveguide and vice versa with low losses ($-2 \, dB$ experimentally demonstrated, $-1 \, dB$ in simulated devices).

9.2.4 Detectors

Wavelength selective detectors are needed in a DWDM NoC to transfer data from the optical domain back to the electronic domain. A natural choice for the detector material is Germanium because it is compatible with CMOS and is well suited to detect light at either 1.3 or 1.55 μm. Germanium detectors that are CMOS compatible and can be integrated with SOI waveguides have been demonstrated by a number of groups. Details can be found in a recent literature review [31]. There are two issues that need to be analyzed regarding the integration of detectors in a DWDM NOC: wavelength-selectivity and power consumption. Detectors demonstrated so far are not wavelength selective; therefore, one would need to build a drop filter. As discussed previously, non-resonant drop filters, such as AWGs, are not well suited for large-scale integration. A better choice is the use of micro-ring resonator drop filters [50] to develop a DWDM link. A more advanced solution would be to integrate the detectors in the drop ring to create a wavelength-specific resonant detector [2]. In both cases, one can arguably build a Ge-on-SOI waveguide detector with capacitance of the order of 10 fF.

Minimizing capacitance is important both to guarantee speed and to limit power consumption [31]. A low-capacitance detector would be able to generate a large voltage drop even for a small optical signal comprising ≃10,000 photons. This would allow one to build a receiver-less detector (i.e., a detector that requires no amplification [31]) or at least limit the number of amplification stages necessary to bring the signal from the detector to the requisite logic level. This would result in both a power and an area savings.

9.2.5 Technology for Intra- and Inter-Chip Communication

In order to compare the performance of optical and electronic on-chip networks, we need a technology road-map for photonics that parallels the ITRS road-map for semiconductors. This estimates the optical device performance at points in time corresponding to future CMOS processes. Although we can expect the performance of electronics to improve with CMOS process scaling, there is no equivalent device scaling effect for photonic devices. Unlike transistors, photonic device geometries do not scale with process feature size since their dimensions are a function of the operating wavelength. For example, the diameter of a micro-ring modulator is constant for a given free spectral range. However, improvements in lithography and manufacturing processes will lead to lower energy solutions. As the manufacturing accuracy of micro-rings improves, less energy is required to tune them. Improvements in detector design reduce detector capacitance and improve the conversion efficiency and result in improved energy efficiency.

When considering the energy per bit of optical communication, it is important to differentiate between simple point to point applications and applications requiring fan-in (many to one) or fan-out (one to many) communication patterns. In a fan-in configuration, there are multiple possible transmitting modulators and one single detector for each wavelength. Even though only one modulator may be active at any one time for a particular wavelength, any inactive micro-ring modulators for that wavelength must be maintained at a resonant wavelength that does not interfere with any of the communication bands. Thus, for high fan-in applications using a single waveguide, the per bit communication energy is highly dependent on the power required to keep inactive modulators tuned.

Our photonic technology road-map assumes the availability of die stacking technology. This allows the use of an unmodified high performance CMOS process for the implementation of the processing elements. Two additional layers are added; an analog layer with the drive electronics and an optical layer containing the modulators, detectors and waveguides. Connection between layers in the stack uses through silicon vias (TSVs) and face to face bonds. Further, CMOS layers dedicated to functions such as local memory can be added depending on the system architecture. The electrical properties of the TSVs limit the number of die-stack layers that will be practical. There is evidence [15] that stack depths of up to eight layers are feasible but thermal issues may further restrict the layer count.

For lower-cost configurations, two layers could be used by integrating the analog electronics with the processor at some cost in area. All the layers in the stack are fabricated on silicon substrates to ensure thermal compliance between layers. The optical layer is relatively simple, consisting of patterning of the optical waveguides and micro-rings, diffusion to create the junctions for the modulators, germanium for the detectors, and a metalization layer to provide contacts between layers. There are no transistors on the optical layer.

An off-chip comb laser operating around 1,310 nm is used to provide optical power. We expect the number of usable wavelengths to increase over time as improvements in manufacturing accuracy allow closer channel spacing. Assuming the use of micro-rings of $5\,\mu$m or less, 64 wavelengths can be accommodated with a channel spacing of 80 GHz. Modulation frequency for the micro-rings will be $10\,\text{Gb s}^{-1}$, giving a total per waveguide bandwidth of $640\,\text{Gb s}^{-1}$.

Our technology roadmap predicts that the energy per bit of photonic communication using this technology will become attractive for NOCs around the time-scale of the 17 nm CMOS process node. At this point, the energy per bit of optical communication is predicted to be less than <85 fJ/bit for point-to-point intra-chip communication and <240 fJ/bit for a fan-in of 64.

9.3 Nanophotonic Network Principles

In this section, we describe basic considerations for photonic NOCs and contrast them with their electrical counterparts.

9.3.1 Electrical Interconnects

Electrical on-chip signals can either use simple repeated *full-swing* wires that are driven between zero and *VDD* volts or through *low-swing* wires that limit the voltage swing to a fraction of *VDD* to save power [32]. We consider these two possibilities in more detail.

The delay of a simple wire is determined by its *RC* time constant (R is resistance, C is capacitance). The resistance and capacitance of a wire of length L are governed by the following equations [20]:

$$R_{wire} = \frac{L\rho}{(thickness - barrier)(width - 2\, barrier)} \tag{9.1}$$

$$C_{wire} = L \left(\varepsilon_0 \left(2K\varepsilon_{horiz}\frac{thickness}{spacing} + 2\varepsilon_{vert}\frac{width}{layerspacing} \right) \right.$$
$$\left. + fringe(\varepsilon_{horiz}, \varepsilon_{vert}) \right) \tag{9.2}$$

Thickness and *width* represent the geometrical dimensions of the wire cross-section, *barrier* represents the thin barrier cladding around the wire to prevent copper from diffusing into the surrounding oxide, and ρ is the material resistivity. The potentially different relative dielectrics for the vertical and horizontal capacitors are represented by ε_{horiz} and ε_{vert}, K accounts for Miller-effect coupling capacitances, *spacing* represents the gap between adjacent wires on the same metal layer, and *layerspacing* represents the gap between adjacent metal layers. Hence, the delay increases quadratically with the length of a wire. A simple technique to overcome this quadratic dependence is to break the wire into multiple smaller segments and connect them with repeaters. As a result, wire delay becomes a linear function of wire length. Overall wire delay can be minimized by selecting proper repeater sizes and spacing between repeaters [6]. Note that these repeaters can be replaced with latches to enable pipelining of the wire in order to boost bandwidth.

Although repeaters reduce delay, they significantly increase energy consumption. As length increases, an optimally repeated wire can consume several times the energy of a simple wire [6]. Energy in the interconnect can be reduced by employing repeaters that are smaller than optimally sized repeaters and by increasing the spacing between successive repeaters. This trades delay for power. The savings will be significant especially in submicron high-performance process technologies where leakage power is very high.

Figure 9.2 shows how deviating from optimal repeater sizing and spacing can impact delay. This figure indicates that if the repeater size is reduced by 50% and the number of repeaters is also reduced by 50%, the delay increases by only 30%. The trade-off between delay and power is wildly nonlinear and depends on many factors. Figure 9.2 addresses repeater size and spacing with respect to delay, but additional

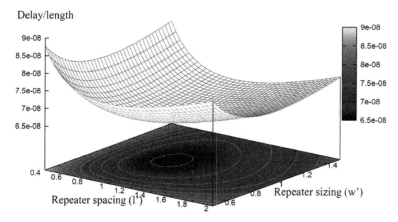

Fig. 9.2   Effect of repeater spacing/sizing on wire delay for global wires at 32 nm technology

factors such as leakage currents, short-circuit currents, process technology, and the physical properties of the chosen wire layer must also be considered to accurately calculate power. The details of this set of influences are too complex to treat properly in this chapter, but an excellent treatment of the topic can be found in [6,33], which show that a $2x$ delay penalty can result in a $5x$ energy savings. The key is that both repeater size and spacing critically influence wire delay and power.

One of the primary reasons for the high power dissipation of global wires is the full-swing (0 V to V_{dd}) requirement imposed by the repeaters. While the repeater spacing and sizing adjustments reduce power to some extent, it remains relatively high. Low-swing alternatives represent another mechanism to vary the wire power, delay, and area trade-off. Reducing the voltage swing on global wires can result in a linear reduction in power. In addition, assuming a separate voltage source for low-swing drivers will result in a quadratic savings in power. However, these large power savings are accompanied by many caveats. Simple receiver-less repeaters do not work with low-swing wires; since we can no longer use repeaters, the delay of a low-swing wire increases quadratically with length. Also, such a wire cannot be pipelined, so it has lower throughput. A low-swing wire requires special transmitter and receiver circuits for signal generation and amplification. This not only increases the area requirement per channel, but also adds extra delay and energy for each bit sent. Due to this fixed overhead, low-swing wires are beneficial only for transferring non-latency-critical data in long on-chip wires.

9.3.2   Optical Interconnect

In an optical interconnect, the power consumed by the optical waveguide itself is very small and is fixed due to the constant power consumed at the laser source.

The overhead primarily lies in the required electric to optical and optical to electrical conversions. Special transmitter and receiver circuits are employed to switch between these domains. Since the load of these circuits is constant and independent of the channel length, their power consumption remains the same for a fixed operating voltage. Hence, unlike other electrical alternatives, the power dissipated by an optical interconnect is relatively independent of the interconnect length. This characteristic makes optics ideal for carrying signals across chip.

A transmitter consists of a waveform generator that is used to tune the ring resonators (discussed in Sect. 9.2.2) into and out of resonance. The power consumed by a transmitter consists of two components: switching energy required to inject carriers into resonators and static energy to maintain the state of the ring. Both the voltage levels of the input signal have to be precisely controlled for proper modulation – the voltage corresponding to logical zero guarantees tuning a ring into resonance, while calibrating the other voltage is critical to avoid inadvertent tuning of ring to a different channel's wavelength. According to Chen et al. [13], the energy required for a logical one to zero transition (tuning ring into resonance) is 14 fJ and the static component at $10\,\mathrm{Gb\,s^{-1}}$ transfer rate is $22\,\mathrm{fJ\,b^{-1}}$. Thus, assuming a 25% probability of a zero to one transition, the overall power consumption of a transmitter is $36\,\mathrm{fJ\,b^{-1}}$.

An optical detector, Sect. 9.2.4, converts an optical signal into electrical current, but the strength of the electrical output from a detector may not be sufficient to drive logical components. A series of amplifiers referred to as a *receiver* are employed to strengthen the weak signal. The detector output is first amplified and converted into voltage by a transimpedance amplifier (TIA). The amplitude of the signal output from a TIA is usually only a few millivolts, and multiple stages of limiting amplifiers are employed to increase the signal strength to VDD.

For a dense deployment of optical interconnects on chip, it is critical to keep the overhead of these amplifiers low. The power consumed by a receiver is determined by the variability of optical components. Due to irregularities in detector ring size and variation in operating temperature, the resulting electrical signal from a detector can deviate significantly from the rated voltage. Depending upon the reliability needs of optical components, the amplifiers in a receiver can either be single ended or differential [40]. A single-ended design uses less resources and has an overhead of 15 fJ/bit. However, faithful amplification of the signal is possible only if the detector output variation is under 10%. A more robust design employs differential amplifiers which take two signal inputs to produce the output voltage. The second input can either be a reference voltage or the complement of the data signal. Compared to a single-ended design, a differential receiver can tolerate up to 75% variation in detector output and still be able to generate valid output. However, the energy cost of a differential receiver is more than twice the cost of the single ended design. Current implementations will likely employ a differential design for reliability. As the technology matures and ring size variation decreases, lower power alternatives will become viable.

9.3.3 Photonic Network Fundamentals

We prefer to arrange the on-chip Si-PIC so that each node has a dedicated, single-destination multiple-source communication channel used for all messages sent to it; we call this arrangement a multiple writer, single reader (MWSR) interconnect, following Pan et al. [38]. In a MWSR, *any* node may write to a given channel but only *one* node (the *destination* or *home* node) may read from the channel.

Optical data packets traverse the channels in a wave-pipelined, or latch-less, manner from a single source to the home node. Packets are fixed in both temporal and spatial extent. Several packets may occupy the same channel simultaneously in moving *slots*. Time of flight varies with source-to-home distances. The MWSR is a common optical interconnect architecture because it deals well with unidirectional wave-pipelined technology [11,30]. In a MWSR, each channel is shared by multiple writers, so an arbiter is necessary to manage write contention.

Some interconnects, Firefly [38] for example, are single writer, multiple reader (SWMR) approaches. Each SWMR channel belongs to *one* sending node, but *any* node may read from the channel. Figure 9.3 contrasts these interconnects and shows how both provide full connectivity. An SWMR benefits from not requiring any arbitration on the part of the sender. The extra SWMR complexity, not present in MWSR, is that the sender must communicate to the receiver that a message is

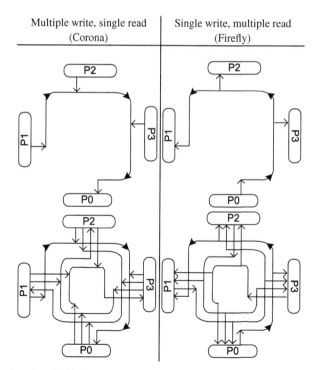

Fig. 9.3 Ring-based optical interconnects

about to be sent. The receiver then activates its detectors, which read the packet and remove it from the waveguide. Firefly broadcasts a head flit to identify the designated receiver of each packet; this costs bandwidth and needs specially designed and relatively expensive broadcast support.

Flow control is needed for an interconnect that can, even for a short time, deliver packets to a receive buffer faster than the destination can drain them and free buffer space. Proposed SWMR designs may experience flow control problems caused by flooding; an SWMR node may receive as many as N flits in a network cycle, where N is the number of interconnected nodes, and this is a hard drain-rate to sustain. An MWSR node can receive at most one flit in a network cycle and as long as it can drain packets at this rate, no flow control is necessary. MWSRs can also combine flow control with channel arbitration, as described in the next section.

In a DWDM system using W wavelengths to connect N nodes, either SWMR or MWSR, N^2W ring resonators are needed. MWSR interconnects require N^2W modulators and NW detectors, while SWMR interconnects require NW modulators and N^2W detectors. To control the growth of the resonator count, one can limit N by using a combined electrical/optical interconnect hierarchy with the electrical interconnect used to connect a local group of nodes to the optical interconnect. In fact, since processor clocks are not increasing, it seems feasible to use electrically connected sub-domains of roughly fixed extent, or perhaps to reduce their size slowly with smaller feature sizes.

9.3.4 Optical Arbitration

In addition to simple data communication, optics can be used to implement certain logical functions such as arbitration [47] and barrier synchronization [8]. In the simplest approach to optical arbitration, presented in Fig. 9.4, a one-bit-wide pulse of monochromatic light travels down an arbitration waveguide. The presence of this light *token* represents the availability of a resource. Each node has a detector on this waveguide. Nodes that need to use the channel activate their detectors (solid- and cross-dotted rings); the other nodes do not activate their detectors (empty dotted rings). At the most one node can detect the token (solid dotted ring), since reading the token removes the light from the waveguide. As a result, a node detecting a token wins exclusive use of the channel. In our protocols, it uses the channel for some fixed period.

Fig. 9.4 Basic optical arbitration

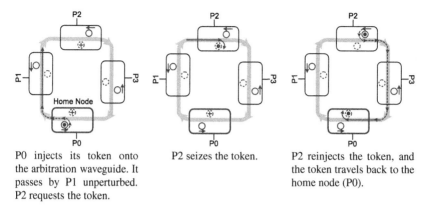

PO injects its token onto the arbitration waveguide. It passes by P1 unperturbed. P2 requests the token.

P2 seizes the token.

P2 reinjects the token, and the token travels back to the home node (P0).

Fig. 9.5 Token channel. Arbitration for one channel of an N channel interconnect

We describe two general protocols: token channel and token slot. Channel protocols allocate the entire channel at once while slot protocols are based on a slotted-ring arbitration protocol [39]. These arbiters may be applied to any optical MWSR, including [25,27,30,46]. In addition, they are general enough to be applied to optical broadcast buses [23,46].

Optical token channel is similar to the 802.5 token ring LAN standard [4]. Figure 9.5 shows the operation of token channel. There is a single token circulated per channel. The token is passed from requester to requester, skipping intervening non-requesters. When a node removes and detects a token, it has exclusive access to the corresponding channel and may begin sending one or more packets some number of cycles later. The sooner it begins transmitting, the better the utilization of the data channel. In token channel, not more than one source can use the channel at any one time: the segment of the data channel from the home node to the token holder carries no data, while the segment from the token holder back to the home node carries light modulated by the token holder. The sender may hold the token, and use the channel, for up to some fixed maximum number $H \geq 1$ of packets. When a sender has many packets in its queue for a particular receiver, channel utilization improves considerably if multiple packets are sent per arbitration round.

When a source first turns on its detector for an otherwise idle channel, it waits on average $T/2$ cycles for arrival of the token. When a single source wants to send a long sequence of packets on an otherwise idle channel, it transmits H packets at a time and waits the full flight time T for the reinjected token to return. The token slot arbiter reduces these token wait times.

The token slot protocol divides the channel into fixed-size, back-to-back slots and circulates tokens in one-to-one correspondence to these slots. One packet occupies one slot. A source waits for arrival of a token, removes the token, and modulates the light in the corresponding single-packet slot. The token precedes the slot by some fixed number of cycles which are used to set up the data for transmission. The slot is occupied by modulated light/data until it gets to the

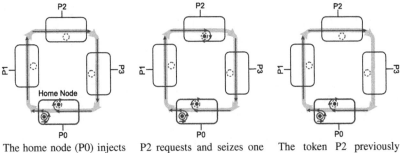

The home node (P0) injects tokens in one-to-one correspondence for the slots.

P2 requests and seizes one of P0's token as it passes by.

The token P2 previously seized is represented by the absence of light traveling by P3.

Fig. 9.6 Token slot

destination. The destination removes the data, frees the slot, and reinjects a fresh token along with new unmodulated light. Figure 9.6 shows the token slot layout and its operation.

Compared with token channel, token slot reduces the average source wait time significantly and increases channel bandwidth utilization in our experiments. In the single-source scenario above, the one active source can claim all tokens, using the channel continuously at full bandwidth. In token channel, the token holder is the only possible sender (there is only one token in the system), whereas in token slot, there are multiple tokens, and there can be multiple sources simultaneously sending at different points on the waveguide.

Unlike token channel, which requires each node to have reinjection capabilities on the arbitration wavelength, only the token slot destination needs reinjection capability. Thus, token slot requires fewer resources and less power, but requires the use of fixed length packets.

This simple token slot protocol is inherently unfair: it gives higher priority to nodes closest to the home. To ensure fairness, we must make sure that every node having sufficiently high demand is treated nearly equally. Details of fairness in both slot and channel protocols are described in [47].

9.3.5 Optical Barriers

In addition to arbitration, low-latency and low-power optical barrier synchronization can be built using nanophotonics [8]. Any barrier function has three components: initialization, arrival, and release. The optical barrier network is a single waveguide, as shown in Fig. 9.7. Each of the N nodes has both a light diverter and a light detector that are attached to the waveguide. There are many possible layouts, but the

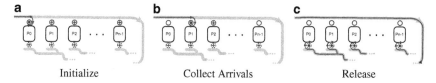

Fig. 9.7 (a) Each participating node begins by diverting the light. No detectors detect light. (b) As participants reach the barrier, diverters cease diverting. Thus far, $P0$ and P_{N-1} have arrived at the barrier. No detectors detect light. (c) All nodes have arrived. As the unicast design shows above, notification is serial, and each node must sequentially detect and then turn off its detector to allow others a chance at the light. In the broadcast design, all nodes quickly detect the light with splitters

key is that the waveguide passes by all nodes twice. In Fig. 9.7, for example, the waveguide wraps around the nodes, with the diverters on one side of the nodes and the detectors on the other.

To initialize the barrier, each participating node diverts light from the waveguide as shown in Fig. 9.7a. Non-participating nodes do not divert the light. As participating nodes arrives at the barrier, they stop diverting the light as shown in Fig. 9.7b. When all nodes have arrived at the barrier (Fig. 9.7c), the light passes by the inactive light diverters and notification begins.

9.4 Corona: A Nanophotonic Case Study

The Corona [46] architecture is a case study that was conducted to evaluate the utility of silicon nanophotonic-based NoCs in the design of a future many-core multiprocessor. Several things are clear: multi- and many-core architectures are **THE** primary focus in today's merchant semiconductor industry, and some media pundits have proclaimed that increasing core counts are the basis for a new "Moore's Law." Unfortunately, the ITRS road-map [42] only predicts a small increase in pin count and pin data rates over the next decade. This implies that the increase in computational capability can somehow be sustained without a commensurate increase in core to core or socket to memory bandwidth. This conclusion is highly suspect unless all application working sets can reside in-cache on the die and that the concurrency profile can be "embarrassingly parallel." Neither is likely.

Several benefits accrue when nanophotonics is coupled to emerging 3D packaging [1]. The 3D approach allows multiple die, each fabricated using a process well suited to it, to be stacked and to communicate with TSVs. Optics, logic, DRAM, non-volatile memory (e.g., FLASH), and analog circuitry may all occupy separate die and co-exist in the same 3D package. Utilizing the third dimension eases layout and helps decrease worst case communication path lengths. 3D packaging is a potential benefit to wire-based interconnect but is even more attractive in systems comprising multiple technologies such as nanophotonic interconnects.

Corona is a nanophotonically connected 3D many-core NUMA system that meets the future bandwidth demands of massively parallel, data-intensive applications at acceptable power levels. Corona is targeted for a 16 nm process in 2017. Corona comprises 256 general purpose cores, organized in 64 four-core clusters, and each cluster is interconnected by an all-optical, high-bandwidth DWDM crossbar. The crossbar enables a cache coherent design with near uniform on-stack and memory communication latencies. Photonic connections to off-stack memory enables unprecedented bandwidth to large amounts of memory with only modest power requirements.

This section describes the Corona architecture and presents a performance comparison to a comparable, all-electrical many-core alternative. The contribution of this work is to show that CMOS compatible nanophotonics is capable of dramatically better communication performance per unit area and energy, and can significantly improve the performance and utility of future many-core architectures in a way that is programmer friendly.

9.4.1 Corona Architecture

Corona is a tightly coupled, highly parallel NUMA system. As NUMA systems and applications scale, it becomes more difficult for the programmer, compiler, and runtime system to manage the placement and migration of programs and data. We try to reduce this burden with homogeneous cores and caches, a crossbar interconnect that has near-uniform latency, a fair interconnect arbitration protocol, and high (one byte per flop) bandwidth between cores and from caches to memory.

The architecture is made up of 256 multi-threaded in-order cores. Corona is capable of supporting up to 1,024 threads simultaneously, providing up to 10 Teraflops of computation, up to 20 TB s^{-1} of on-stack bandwidth, and up to 10 TB s^{-1} of off-stack memory bandwidth. Figure 9.8 gives a conceptual view of the system while Fig. 9.9 provides a sample layout of the system including the waveguides that comprise the optical interconnect, the optical connection to memory, and other optical components.

Each core has a private L1 instruction and data cache, and all four cores share a unified L2 cache. A hub routes message traffic between the L2, directory, memory controller, network interface, optical bus, and optical crossbar. Figure 9.8b shows the cluster configuration, while the upper left-hand insert in Fig. 9.9 shows its approximate floor-plan.

Because Corona is an architecture targeting future high throughput systems, our exploration and evaluation of the architecture are *not* targeted at optimal subcomponent configuration (such as their branch prediction schemes, number of execution units, cache sizes, and cache policies). Rather, the design decisions shown in Table 9.1 represent reasonably modest choices for a high-performance system targeted at a 16-nm process in 2017.

Fig. 9.8 (a) Architecture overview, (b) Cluster detail

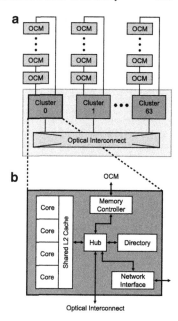

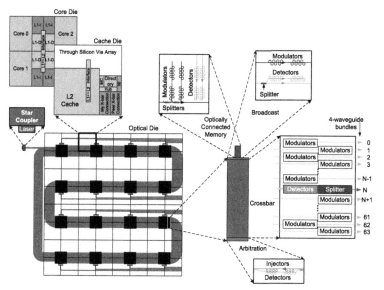

Fig. 9.9 Layout with serpentine crossbar and resonator ring detail

Power analysis has been based on the Penryn [22] and the Silverthorne [21] cores. Penryn is a single-threaded out-of-order core supporting 128-bit SSE4 instructions. Power per core has been conservatively reduced by 5× (compared to the 6× predictions in [5]) and then increased by 20% to account for differences in the quad-threaded Corona. Silverthorne is a dual-threaded in-order 64-bit design

Table 9.1 Resource
configuration

Resource	Value
Number of clusters	64
Per-cluster:	
L2 cache size/assoc	4 MB/16-way
L2 cache line size	64 B
L2 coherence	MOESI
Memory controllers	1
Cores	4
Per-core:	
L1 I cache size/assoc	16 KB/4-way
L1 D cache size/assoc	32 KB/4-way
L1 I & D cache line size	64 B
Frequency	5 GHz
Threads	4
Issue policy	In-order
Issue width	2
64 b floating point SIMD width	4
Fused floating point operations	Multiply-add

where power and area have been similarly increased to account for the Corona architectural parameters. Directory and L2 cache power have been calculated using CACTI 5 [44]. Hub and memory controller power estimates are based on synthesized 65 nm designs and synopsis nanosim power values scaled to 16 nm. Total processor, cache, memory controller, and hub power for the Corona design are expected to be between 82 W (Silverthorne based) and 155 W (Penryn based).

Area estimates are based on pessimistically scaled Penryn and Silverthorne designs. We estimate that an in-order Penryn core will have one-third the area of the existing out-of-order Penryn. This estimate is consistent with the current core-size differences in 45 nm for the out-of-order Penryn and the in-order Silverthorne, and is more conservative than the $5x$ area reduction reported by Asanovic et al. [5]. We then assume a multi-threading area overhead of 10% as reported in Chaudry et al. [12]. Total die area for the processor and L1 die is estimated to be between 423 mm^2 (Penryn based) and 491 mm^2 (Silverthorne based). The discrepancy between these estimates is affected by the six-transistor Penryn L1 cache cell design vs. the eight-transistor Silverthorne L1 cache cell.

The Corona architecture has one memory controller per cluster. Associating the memory controller with the cluster ensures that the memory bandwidth grows linearly with increased core count, and it provides local memory accessible with low latency. Photonics connects the memory controller to off-stack main memory. Network interfaces, similar to the interface to off-stack main memory, provide inter-stack communication for larger systems using DWDM interconnects. Corona's 64 clusters communicate through an optical crossbar and occasionally an optical broadcast ring. Both are managed using the token channel protocol. Several messages of different sizes may simultaneously share any communication channel, allowing for high utilization.

Table 9.2 Optical resource inventory

Photonic subsystem	Waveguides	Ring resonators
Memory	128	16 K
Crossbar	256	1,024 K
Broadcast	1	8 K
Arbitration	2	8 K
Clock	1	64
Total	388	≈1,056 K

Table 9.2 summarizes the NoC's optical component requirements (power waveguides and I/O components are omitted). Based on existing designs, we estimate that the photonic interconnect power (including the power dissipated in the analog circuit layer and the laser power in the photonic die) to be 39 W.

Each cluster has a designated channel that address, data, and coherence messages share. Any cluster may write to a given channel, but only a single fixed cluster may read from the channel. A fully connected 64×64 crossbar can be realized by replicating this many-writer single-reader channel 64 times, adjusting the assigned "reader" cluster with each replication.

The channels comprise 256 wavelengths, or four bundled waveguides each having 64 DWDM wavelengths and use a MWSR approach. Light is sourced at a channel's home by a splitter that provides all wavelengths of light from a power waveguide. Communication is unidirectional, in cyclically increasing order of cluster number.

A cluster sends to another cluster by modulating the light on the destination cluster's channel. Figure 9.10 illustrates the conceptual operation of a four-wavelength channel. Modulation occurs on both clock edges, so that each of the wavelengths signals at $10 \, \text{Gb s}^{-1}$, yielding a per-cluster bandwidth of $2.56 \, \text{Tb s}^{-1}$ and a total crossbar bandwidth of $20 \, \text{TB s}^{-1}$.

A wide phit with low modulation time minimizes packet latency and in-order core stall time. A 64-byte cache line can be sent (256 bits in parallel twice per clock) in one 5 GHz clock. The propagation time is at most eight clocks and is determined by a combination of the source's distance from the destination and the speed of light in a silicon waveguide (approximately 2 cm per clock). Because messages, such as cache lines, are localized to a small portion of the bundle's length, a bundle may have multiple back-to-back messages in transit simultaneously.

Corona uses optical global distribution of the clock in order to avoid the need for signal re-timing at the destination. A clock distribution waveguide parallels the data waveguide, with the clock signal traveling clockwise with the data signals. This means that each cluster is offset from the previous cluster by approximately one-eighth of a clock cycle. A cluster's electrical clock is phase locked to the arriving optical clock. Thus, input and output data are in phase with the local clock; this avoids costly re-timing except when the serpentine wraps around.

A key design goal is to scale main memory bandwidth to match the growth in computational power. Maintaining this balance ensures that the performance of the system is not overly dependent on the cache access patterns of the application.

Fig. 9.10 A four wavelength data channel example. The home cluster (cluster 1) sources all wavelengths of light (r,g,b,y). The light travels clockwise around the crossbar waveguides. It passes untouched by cluster 2s inactive (off-resonance) modulators. As it passes by cluster 3s active modulators, all wavelengths are modulated to encode data. Eventually, cluster 1s detectors sense the modulation, at which point the waveguide terminate

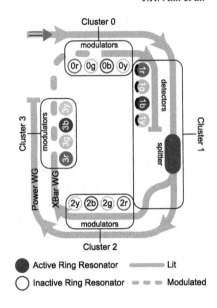

Our target external memory bandwidth for a 10 Teraflops processor is $10\,\mathrm{TB\,s^{-1}}$. Using an electrical interconnect to achieve this performance would require excessive power; over 160 W assuming $2\,\mathrm{mW\,Gb^{-1}\,s^{-1}}$ [37] interconnect power. Instead, we use a nanophotonic interconnect that has high bandwidth and low power. The same channel separations and data rates that are used on the internal interconnect network can also be used for external fiber connections. We estimate the interconnect power to be $0.078\,\mathrm{mW\,Gb^{-1}\,s^{-1}}$, which equates to a total memory system interconnect power of approximately 6.4 W.

Each of the 64 memory controllers connects to its external memory by a pair of single-waveguide, 64-wavelength DWDM links. The optical network is modulated on both edges of the clock. Hence, each memory controller provides $160\,\mathrm{GB\,s^{-1}}$ of off-stack memory bandwidth, and all memory controllers together provide $10\,\mathrm{TB\,s^{-1}}$.

This allows all communication to be scheduled by the memory controller with no arbitration. Each external optical communication link consists of a pair of fibers providing half-duplex communication between the CPU and a string of optically connected memory (OCM) modules. The link is optically powered from the chip stack; after connecting to the OCMs, each outward fiber is looped back as a return fiber. Although the off-stack memory interconnect uses the same modulators and detectors as the on-stack interconnects, the communication protocols differ. Communication between processor and memory is master/slave, as opposed to peer-to-peer. To transmit, the memory controller modulates the light and the target module diverts a portion of the light to its detectors. To receive, the memory controller detects light that the transmitting OCM has modulated on the return fiber. Because the memory controller is the master, it can supply the necessary unmodulated power to the transmitting OCM.

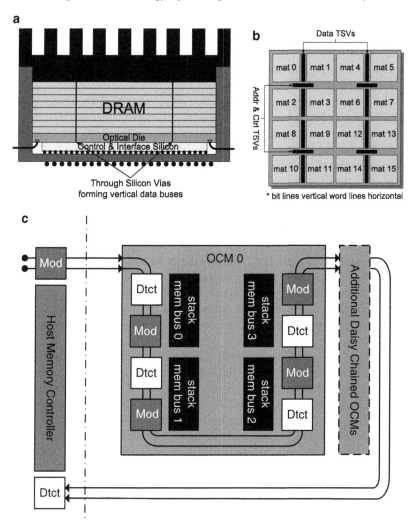

Fig. 9.11 Schematic view of optically connected memory. (**a**) 3D die stack. The stack has one optical die and multiple DRAM dies. (**b**) DRAM die floor-plan. Each quadrant is independent, and could also be constructed from four independent die. (**c**) OCM expansion. The light travels from the processor, through one or more OCMs, finally looping back to the processor

Figure 9.11a shows the 3D stacked OCM module, built from custom DRAM die and an optical die. The DRAM die is organized so that an entire cache line is read or written from a single mat. 3D stacking is used to minimize the delay and power in the interconnect between the optical fiber loop and the DRAM mats. The high-performance optical interconnect allows a single mat to quickly provide all the data for an entire cache line. In contrast, current electrical memory systems and DRAMs activate many banks on many die on a DIMM, reading out tens of thousands of bits into an open page. However, with highly interleaved memory systems and a thousand threads, the chances of the next access being to an open page are small. Corona's DRAM architecture avoids accessing an order of magnitude more bits than are needed for the cache line, and hence consumes less power in its memory system.

Corona supports memory expansion by adding additional OCMs to the fiber loop as shown in Fig. 9.11c. Expansion adds only modulators and detectors and not lasers; so the incremental communication power is small. As the light passes directly through the OCM without buffering or re-timing, the incremental delay is also small, so that the memory access latency is similar across all modules. In contrast, a serial electrical scheme, such as FBDIMM, would typically require the data to be re-sampled and retransmitted at each module, increasing the communication power and access latency.

Figure 9.12 illustrates the Corona 3D die stack. Most of the signal activity, and therefore heat, are in the top die (adjacent to the heat sink), which contains the clustered cores and L1 caches. The processor die is face-to-face bonded with the L2 die, providing direct connection between each cluster and its L2 cache, hub, memory controller, and directory. The bottom die contains all of the optical structures (waveguides, ring resonators, detectors, etc.) and is face-to-face bonded with the analog electronics which contain detector circuits and control ring resonance and modulation.

All of the L2 die components are potential optical communication end points and connect to the analog die by signal through silicon TSVs. This strategy minimizes the layout impact since most die-to-die signals are carried in the face-to-face bonds. External power, ground, and clock TSVs are the only TSVs that must go through three die to connect the package to the top two digital layers. The optical die is larger than the other die in order to expose a *mezzanine* to permit fiber attachments for I/O and OCM channels and external lasers.

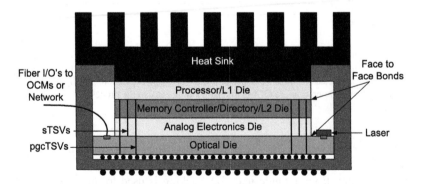

Fig. 9.12 Schematic side view of 3D package

9.4.2   Experimental Setup

The Corona architecture was subjected to a combination of synthetic and realistic workloads that were selected with an eye to stressing the on-stack and memory interconnects. Synthetic workloads stress particular features and aspects of the interconnects. The SPLASH-2 benchmark suite [48] indicates their realistic performance. The SPLASH-2 applications are not modified in their essentials. Larger datasets were used, when possible, to ensure that each core has a nontrivial workload. Because of a limitation in our simulator, we needed to replace implicit synchronization via semaphore variables with explicit synchronization constructs. In addition, we set the L2 cache size to 256 kb to better match our simulated benchmark size and duration when scaled to expected system workloads. A summary of the workload setup is described in Table 9.3.

The simulation infrastructure is split into two independent parts: a full system simulator for generating L2 miss memory traces and a network simulator for processing these traces. A modified version of the HP Labs' COTSon simulator [16] generates the traces. COTSon is based on AMD's SimNow simulator infrastructure. Each application is compiled with gcc 4.1, using -O3 optimization, and run as a single 1,024-threaded instance. We are able to collect multi-threaded traces by translating the operating system's thread-level parallelism into hardware thread-level

Table 9.3 Benchmarks and configurations

Synthetic		
Benchmark	Description	# Network requests
Uniform	Uniform random	1 M
Hot spot	All clusters to one cluster	1 M
Tornado	Cluster (i, j) to cluster $((i + \lfloor k/2 \rfloor - 1)\%k, (j + \lfloor k/2 \rfloor - 1)\%k)$, where k = network's radix	1 M
Transpose	Cluster (i, j) to cluster (j, i)	1 M

SPLASH-2			
Benchmark	Data set Experimental (default)		# Network requests
Barnes	64 K particles	(16 K)	7.2 M
Cholesky	tk29.O	(tk15.O)	0.6 M
FFT	16 M points	(64 K)	176 M
FMM	1 M particles	(16 K)	1.8 M
LU	2,048×2,048 matrix	(512×512)	34 M
Ocean	2,050×2,050 grid	(258×258)	240 M
Radiosity	Roomlarge	(room)	4.2 M
Radix	64 M integers	(1 M)	189 M
Raytrace	Balls4	(car)	0.7 M
Volrend	Head	(head)	3.6 M
Water-Sp	32 K molecules	(512)	3.2 M

parallelism. In order to keep the trace files and network simulations manageable, the simulators do not tackle the intricacies of cache coherency between clusters.

The network simulator reads the traces and processes them in the network subsystem. The traces consist of L2 misses and synchronization events that are annotated with thread ID and timing information. In the network simulator, L2 misses go through a request-response, on-stack interconnect transaction and an off-stack memory transaction. The simulator, which is based on the M5 framework [9], takes a trace-driven approach to processing memory requests. The MSHRs, hub, interconnect, arbitration, and memory are all modeled in detail with finite buffers, queues, and ports. This enforces bandwidth, latency, back pressure, and capacity limits throughout.

In the simulation, our chief goal is to understand the performance implications of the on-stack network and the off-stack memory design. Our simulator has three network configuration options:

- XBar – An optical crossbar (as described in the previous section), with bisection bandwidth of $20.48\,\text{TB s}^{-1}$ and maximum signal propagation time of eight clocks.
- HMesh – An electrical 2D mesh with bisection bandwidth $1.28\,\text{TB s}^{-1}$ and per hop signal latency (including forwarding and signal propagation time) of five clocks.
- LMesh – An electrical 2D mesh with bisection bandwidth $0.64\,\text{TB s}^{-1}$ and per hop signal latency (including forwarding and signal propagation time) of five clocks.

The two meshes employ dimension-order worm-hole routing [14]. Since many components of the optical system power are fixed (e.g., laser, ring trimming, etc.), we conservatively estimated a continuous power of 26 W for the XBar. We assumed an electrical energy of 196 pJ per transaction per hop, including router overhead. These assumptions are based on aggressively scaled low-swing busses and ignores all leakage power in the electrical meshes.

We also simulate two memory interconnects, the OCM interconnect of the section above and an electrical interconnect to memory:

- OCM – Optically connected memory; off-stack memory bandwidth is $10.24\,\text{TB s}^{-1}$, memory latency is 20 ns.
- ECM – Electrically connected memory; off-stack memory bandwidth is $0.96\,\text{TB s}^{-1}$, memory latency is 20 ns.

The electrical memory interconnect is based on the ITRS road-map, which indicates that it will be impossible to implement an ECM with performance equivalent to the proposed OCM. Table 9.4 contrasts the memory interconnects.

We simulate five combinations: XBar/OCM (i.e., Corona), HMesh/OCM, LMesh-/OCM, HMesh/ECM, and LMesh/ECM. These choices highlight, for each benchmark, the performance gain, if any, faster memory and due to faster interconnect. We ran each simulation for a predetermined number of network requests (L2 misses). These miss counts are shown in Table 9.3.

Table 9.4 Optical vs Electrical Memory Interconnects	Resource	OCM	ECM
	Memory controllers	64	64
	External connectivity	256 fibers	1,536 pins
	Channel width	128 b half duplex	12 b full duplex
	Channel data rate	$10\,\mathrm{Gb\,s^{-1}}$	$10\,\mathrm{Gb\,s^{-1}}$
	Memory bandwidth	$10.24\,\mathrm{TB\,s^{-1}}$	$0.96\,\mathrm{TB\,s^{-1}}$
	Memory latency	20 ns	20 ns

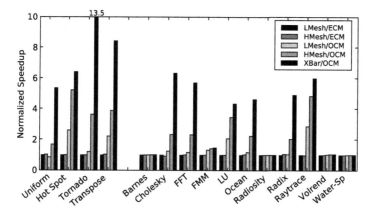

Fig. 9.13 Normalized speedup

9.4.3 Performance Evaluation

For the five system configurations, Fig. 9.13 shows performance relative to the realistic, electrically connected LMesh/ECM system.

We can form a few hypotheses based on the synthetic benchmarks. Understandably, with low memory bandwidth, the high-performance mesh adds little value. With fast OCM, there is very substantial performance gain over ECM systems, when using the fast mesh or the crossbar interconnect, but much less gain if the low performance mesh is used. Most of the performance gain made possible by OCM is realized only if the crossbar interconnect is used. In the exceptional case, Hot Spot, memory bandwidth remains the performance limiter (because all the memory traffic is channeled through a single cluster); hence, there is less pressure on the interconnect. Overall, by moving to an OCM from an ECM in systems with an HMesh, we achieve a geometric mean speedup of 3.28. Adding the photonic crossbar can provide a further speedup of 2.36 on the synthetic benchmarks.

For the SPLASH-2 applications, we find that in four cases (Barnes, Radiosity, Volrend, and Water-Sp) the LMesh/ECM system is fully adequate. These applications perform well due to their low cache-miss rates and consequently low main memory bandwidth demands. FMM is quite similar to these. The remaining applications are memory bandwidth limited on ECM-based systems. For Cholesky, FFT, Ocean, and Radix, fast memory provides considerable benefits, which are

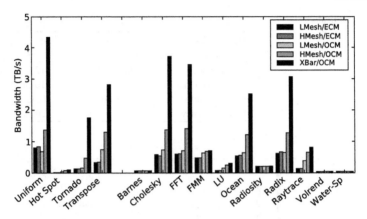

Fig. 9.14 Achieved bandwidth

realized only with the fast crossbar. LU and Raytrace are like Hot Spot: while OCM provides the most significant speedup, some additional benefit is derived from the fast crossbar. We posit below a possible reason for the difference between Cholesky, FFT, Ocean, and Radix on the one hand, and LU and Raytrace on the other hand, when examining the bandwidth and latency data. These observations are generally consistent with the detailed memory traffic measurements reported by Woo et al. [48]. Overall, replacing an ECM with an OCM in a system using an HMesh can provide a geometric mean speedup of 1.80. Adding the photonic crossbar can provide a further speedup of 1.44 on the SPLASH-2 applications.

Figure 9.14 shows the actual rate of communication with main memory. Figure 9.14 shows that the four low bandwidth applications that perform well on the LMesh/ECM configuration are those with bandwidth demands lower than that provided by ECM. FMM needs somewhat more memory bandwidth than ECM provides. Three of the synthetic tests and four of the applications have very high bandwidth and interconnect requirements in the 2–5 TB s^{-1} range; these benefit the most from the XBar/OCM configuration. LU and Raytrace do much better on OCM systems than ECM but do not require much more bandwidth than ECM provides. They appear to benefit mainly from the improved latency offered by XBar/OCM. This is due to bursty memory traffic in these two applications. Analysis of the LU code shows that many threads attempt to access the same remotely stored matrix block at the same time, following a barrier. In a mesh, this over-subscribes the links into the cluster that stores the requested block.

Figure 9.15 reports the average latency of an L2 cache miss to main memory. An L2 miss may be delayed in waiting for the arbitration token and by destination flow-control before a miss-service message can be sent. Our latency statistics measure both queue waiting times and interconnect transit times. LU and Raytrace see considerable average latency in ECM systems; it is improved dramatically by OCM and improved further by the optical crossbar. Note that the average latency can be high even when overall bandwidth is low when traffic is bursty.

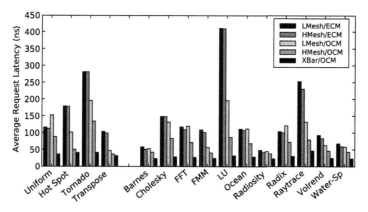

Fig. 9.15 Average L2 miss latency

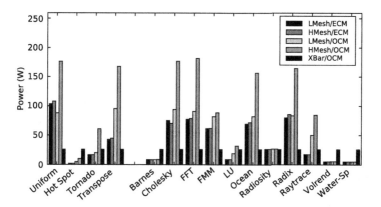

Fig. 9.16 On-chip network power

Figure 9.16 shows the on-chip network dynamic power. For applications that fit in the L2 cache, the photonic crossbar can dissipate more power than for the electronic meshes, albeit ignoring mesh leakage power. However, for applications with significant memory demands, the power of the electronic meshes can fast become prohibitive with power of 100 W or more, while providing lower performance.

9.5 Summary

The new Moore's Law for cores will place tremendous stresses on the bandwidth achievable and the power consumed by electrical chip-wide and inter-chip interconnection networks. Pin limitations and the power and latency issue of long wires are the chief contributors to the problem. To solve this problem, the technology, topology, and protocols of NOCs, processor to memory interconnect, and memory structures will need to be re-examined. Current research, ours and that of other investigators, has shown that nanophotonics can have a dramatic impact on energy, cost, and performance, and that this will need to be done through a significant re-architecting of the interconnection architecture of the chip multiprocessor.

In this chapter, we have presented the fundamental properties of both wire-based and CMOS compatible nanophotonic NOCs. The presentation has been grounded with a reasonably detailed discussion of wire properties and the photonic components that provide an opportunity to replace electrical NOCs with photonic NOCs that may solve show-stopping problems that electrical NOCs will encounter in the many-core processors of the future. The Corona case study was used to compare the tradeoffs between wire and waveguide-based interconnects in a machine architecture that could potentially exist in the 2017 time frame if chip multiprocessor core counts continue to increase under a Moore's Law trajectory.

We recognize that it is a very long way from laboratory device demonstrations of today to reliable, cheap, and high-yield manufacture of nanophotonic structures that will be necessary to make Corona-like devices a reality. But systems with Corona-like performance will be required in order to continue the growth of computation power that is needed. Clearly, these systems will only achieve their performance promise if on-chip inter-core interconnect bandwidth can scale with the increase in core count and improved performance per core, and if future main memory bandwidth can scale at a rate commensurate with the performance of the many-core processor socket. It remains to be seen how these aggressive goals will be achieved in commercial practice. We contend that it is unlikely that they can be achieved using evolved electrical NOC technology within feasible power and thermal constraints. New technology approaches are needed. We also contend that from the today's perspective, the technology with the greatest promise to meet these needs is silicon nanophotonics.

References

1. Proceedings of the ISSCC Workshop F2: Design of 3D-Chipstacks. IEEE (2007). Organizers: W. Weber and W. Bowhill
2. Ahn, J., Fiorentino, M., Beausoleil, R., Binkert, N., Davis, A., Fattal, D., Jouppi, N., McLaren, M., Santori, C., Schreiber, R., Spillane, S., Vantrease, D., Xu, Q.: Devices and architectures for photonic chip-scale integration. Applied Physics A: Materials Science and Processing **95**(4), 989–997 (2009)
3. Analui, B., Guckenberger, D., Kucharski, D., Narasimha, A.: A fully integrated 20-gb/s opto-electronic transceiver implemented in a standard 0.13 μm cmos soi technology. IEEE Journal of Solid-State Circuits **41**(12), 2945–2955 (2006)
4. ANSI/IEEE: Local Area Networks: Token Ring Access Method and Physical Layer Specifications, Std 802.5. Tech. rep. (1989)
5. Asanovic, K., Bodik, R., Catanzaro, B.C., Gebis, J.J., Husbands, P., Keutzer, K., Patterson, D.A., Plishker, W.L., Shalf, J., Williams, S.W., Yelick, K.A.: The Landscape of Parallel Computing Research: A View from Berkeley. Tech. Rep. UCB/EECS-2006-183, EECS Department, University of California, Berkeley (2006)
6. Banerjee, K., Mehrotra, A.: A power-optimal repeater insertion methodology for global interconnects in nanometer designs. IEEE Transactions on Electron Devices **49**(11), 2001–2007 (2002)

7. Batten, C., Joshi, A., Orcutt, J., Khilo, A., Moss, B., Holzwarth, C., Popvic, M., Li, H., Smitth, H., Hoyt, J., Kartner, F., Ram, R., Stojanovic, V., Asanovic, K.: Building manycore processor-to-DRAM networks with monolithic silicon photonics. In: Hot Interconnects (2008)
8. Binkert, N., Davis, A., Lipasti, M., Schreiber, R., Vantrease, D.: Nanophotonic Barriers. In: Workshop on Photonic Interconnects and Computer Architecture (2009)
9. Binkert, N.L., Dreslinski, R.G., Hsu, L.R., Lim, K.T., Saidi, A.G., Reinhardt, S.K.: The M5 simulator: Modeling networked systems. IEEE Micro 26(4), 52–60 (2006)
10. Black, B., et al: Die Stacking 3D Microarchitecture. In: Proceedings of MICRO-39. IEEE (2006)
11. Bogineni, K., Sivalingam, K.M., Dowd, P.W.: Low-complexity multiple access protocols for wavelength-division multiplexed photonic networks. IEEE Journal on Selected Areas in Communications 11, 590–604 (1993)
12. Chaudhry, S., Caprioli, P., Yip, S., Tremblay, M.: High-performance throughput computing. IEEE Micro 25(3), 0272–1732 (2005)
13. Chen, L., Preston, K., Manipatruni, S., Lipson, M.: Integrated GHz silicon photonic interconnect with micrometer-scale modulators and detectors. Optical Express 17, 15248–15256 (2009)
14. Dally, W., Seitz, C.: Deadlock-free message routing in multiprocessor interconnection networks. IEEE Transactions on Computers C-36(5), 547–553 (1987)
15. Davis, W.R., Wilson, J., Mick, S., Xu, J., Hua, H., Mineo, C., Sule, A., Steer, M., Franzon, P.: Demystifying 3D ICS: The pros and cons of going vertical. IEEE Design and Test of Computers 22(1), 498–510 (2005)
16. Falcon, A., Faraboschi, P., Ortega, D.: Combining Simulation and Virtualization through Dynamic Sampling. In: ISPASS (2007)
17. Fischer, U., Zinke, T., Kropp, J.R., Arndt, F., Petermann, K.: 0.1 db/cm waveguide losses in single-mode soi rib waveguides. IEEE Photonics Technology Letters 8(5), 647–648 (1996)
18. Green, W.M., Rooks, M.J., Sekaric, L., Vlasov, Y.A.: Ultra-compact, low rf power, 10 gb/s siliconmach-zehnder modulator. Optical Express 15(25), 17106–17113 (2007)
19. Gubenko, A., Krestnikov, I., Livshtis, D., Mikhrin, S., Kovsh, A., West, L., Bornholdt, C., Grote, N., Zhukov, A.: Error-free 10 gbit/s transmission using individual fabry-perot modes of low-noise quantum-dot laser. Electronics Letters 43(25), 1430–1431 (2007)
20. Ho, R., Mai, K., Horowitz, M.: The Future of Wires. Proceedings of the IEEE, Vol. 89, No. 4 (2001)
21. Intel: Intel Atom Processor. http://www.intel.com/techno-logy/atom
22. Intel: Introducing the 45nm Next Generation Intel Core Microarchitecture. http://www.intel.com/technology/magazine/ 45nm/coremicroarchitecture-0507.htm
23. Kirman, N., Kirman, M., Dokania, R.K., Martinez, J.F., Apsel, A.B., Watkins, M.A., Albonesi, D.H.: Leveraging Optical Technology in Future Bus-based Chip Multiprocessors. In: MICRO'06, pp. 492–503. IEEE Computer Society, Washington, DC, USA (2006)
24. Koch, B.R., Fang, A.W., Cohen, O., Bowers, J.E.: Mode-locked silicon evanescent lasers. Optics Express 15(18), 11225 (2007)
25. Kodi, A., Louri, A.: Performance adaptive power-aware reconfigurable optical interconnects for high-performance computing (hpc) systems. In: SC '07: Proceedings of the 2007 ACM/IEEE conference on Supercomputing, pp. 1–12. ACM, NY, USA (2007)
26. Kovsh, A., Krestnikov, I., Livshits, D., Mikhrin, S., Weimert, J., Zhukov, A.: Quantum dot laser with 75nm broad spectrum of emission. Optics Letters 32(7), 793–795 (2007)
27. Krishnamurthy, P., Franklin, M., Chamberlain, R.: Dynamic reconfiguration of an optical interconnect. In: ANSS '03, p. 89. IEEE Computer Society, Washington, DC, USA (2003)
28. Kumar, R., Zyuban, V., Tullsen, D.M.: Interconnections in Multi-Core Architectures: Understanding Mechanisms, Overheads and Scaling. In: ISCA-32, pp. 408–419. IEEE Computer Society, Washington, DC, USA (2005)
29. Lipson, M.: Guiding, modulating, and emitting light on silicon–challenges and opportunities. Journal of Lightwave Technology 23(12), 4222–4238 (2005)
30. Marsan, M.A., Bianco, A., Leonardi, E., Morabito, A., Neri, F.: All-optical WDM multi-rings with differentiated QoS. IEEE Communications Magazine 37(2), 58–66 (1999)

31. Miller, D.: Device requirements for optical interconnects to silicon chips. Proceedings of the IEEE **97**(7), 1166–1185 (2009)
32. Muralimanohar, N.: Interconnect Aware Cache Architectures. Ph.D. thesis, University of Utah (2009)
33. Muralimanohar, N., Balasubramonian, R., Jouppi, N.: Optimizing NUCA Organizations and Wiring Alternatives for Large Caches with CACTI 6.0. In: Proceedings of the 40th International Symposium on Microarchitecture (MICRO-40) (2007)
34. Nagarajan, R., et al: Large-scale photonic integrated circuits for long-haul transmission and switching. Journal of Optical Networking **6**(2), 102–111 (2007)
35. Nawrocka, M., Tao Liu, Xuan Wang, Panepucci, R.: Tunable silicon microring resonator with wide free spectral range. Applied Physics Letters **89**(7), 071110 (2006)
36. Owens, J., Dally, W., Ho, R., Jayasimha, D., Keckler, S., Peh, L.S.: Research challenges for on-chip interconnection networks. IEEE Micro **27**(5), 96–108 (2007)
37. Palmer, R., Poulton, J., Dally, W.J., Eyles, J., Fuller, A.M., Greer, T., Horowitz, M., Kellam, M., Quan, F., Zarkeshvarl, F.: A 14mW 6.25Gb/s Transceiver in 90nm CMOS for Serial Chip-to-Chip Communications. In: ISSCC (2007)
38. Pan, Y., Kumar, P., Kim, J., Memik, G., Zhang, Y., Choudhary, A.: Firefly: illuminating future network-on-chip with nanophotonics. In: ISCA '09, pp. 429–440. ACM, NY, USA (2009)
39. Pierce, J.: How far can data loops go? IEEE Transactions on Communications **20**(3), 527–530 (1972)
40. Razavi, B.: Design of Integrated Circuits for Optical Communications. McGraw-Hill, NY (2003)
41. Roelkens, G., Vermeulen, D., Thourhout, D.V., Baets, R., Brision, S., Lyan, P., Gautier, P., Fédéli, J.M.: High efficiency diffractive grating couplers for interfacing a single mode optical fiber with a nanophotonic silicon-on-insulator waveguide circuit. Applied Physics Letters **92**(13), 131101 (2008)
42. Semiconductor Industries Association: International Technology Roadmap for Semiconductors. http://www.itrs.net/ (2006 Update)
43. Shijun Xiao, Khan, M., Hao Shen, Minghao Qi: A highly compact third-order silicon microring add-drop filter with a very large free spectral range, a flat passband and a low delay dispersion. Optics Express **15**(22), 14,765–71 (2007)
44. Thoziyoor, S., Muralimanohar, N., Ahn, J., Jouppi, N.P.: CACTI 5.1. Tech. Rep. HPL-2008-20, HP Labs
45. Trotter, M.R.W.D.C., Young, R.W.: Maximally Confined High-Speed Second-Order Silicon Microdisk Switches. In: OSA Technical Digest (2008)
46. Vantrease, D., Schreiber, R., Monchiero, M., McLaren, M., Jouppi, N.P., Fiorentino, M., Davis, A., Binkert, N., Beausoleil, R.G., Ahn, J.: Corona: System Implications of Emerging Nanophotonic Technology. In: ISCA-35, pp. 153–164 (2008)
47. Vantrease, D., Binkert, N., Schreiber, R., Lipasti, M.H.: Light speed arbitration and flow control for nanophotonic interconnects. In: Micro-42: Proceedings of the 42nd Annual IEEE/ACM International Symposium on Microarchitecture, pp. 304–315. ACM, NY, USA (2009)
48. Woo, S.C., Ohara, M., Torrie, E., Singh, J.P., Gupta, A.: The SPLASH-2 Programs: Characterization and Methodological Considerations. In: ISCA, pp. 24–36 (1995)
49. Xu, Q., Schmidt, B., Pradhan, S., Lipson, M.: Micrometre-scale silicon electro-optic modulator. Nature **435**(7040), 325–327 (2005)
50. Xu, Q., Schmidt, B., Shakya, J., Lipson, M.: Cascaded silicon micro-ring modulators for wdm optical interconnection. Optical Express **14**(20), 9431–9435 (2006)
51. Xu, Q., Fattal, D., Beausoleil, R.G.: Silicon microring resonators with 1.5 μm radius. Optical Express **16**(6), 4309–4315 (2008)
52. Xue, J., et al: An Intra-Chip Free-Space Optical Interconnect. In: CMP-MSI: 3rd Workshop on Chip Multiprocessor Memory Systems and Interconnects (2008)
53. Yu-Hsuan Kuo, Yong Kyu Lee, Yangsi Ge, Shen Ren, Roth, J., Kamins, T., Miller, D., Harris, J.: Strong quantum-confined Stark effect in germanium quantum-well structures on silicon. Nature **437**(7063), 1334–6 (2005)

Chapter 10
RF-Interconnect for Future Network-On-Chip

Sai-Wang Tam, Eran Socher, Mau-Chung Frank Chang, Jason Cong, and Glenn D. Reinman

Abstract In the era of the nanometer CMOS technology, due to stringent system requirements in power and performance, microprocessor manufacturers are relying more on chip multi-processor (CMP) designs. CMPs partition silicon real estate among a number of processor cores and on-chip caches, and these components are connected via an on-chip interconnection network (Network-on-chip). It is projected that communication via NoC is one of the primary limiters to both performance and power consumption. To mitigate such problems, we explore the use of multiband RF-interconnect (RF-I) which can communicate simultaneously through multiple frequency bands with low power signal transmission and reconfigurable bandwidth. At the same time, we investigate the CMOS mixed-signal circuit implementation challenges for improving the RF-I signaling integrity and efficiency. Furthermore, we propose a micro-architectural framework that can be used to facilitate the exploration of scalable low power NoC architectures based on physical planning and prototyping.

10.1 Introduction

In the era of the nanometer CMOS technology, due to stringent system requirements in power and performance, processor manufacturers are relying more on chip multi-processor (CMP) designs instead of single-core design with high clocking frequency and deep pipelining architecture. Recent studies [1] also project that heterogeneous many-core designs with massive parallel data processing and distributed caches will be the dominant mobile system architecture to satisfy future application needs. However, many-core computation requires the partitioning of silicon real estate among a large number of processor cores and memory caches. As a result, the

S.-W. Tam (✉)
Electrical Engineering Department, University of California, Los Angeles,
Engineering IV Building, Los Angeles, CA 90095, USA
e-mail: roccotam@ee.ucla.edu

C. Silvano et al. (eds.), *Low Power Networks-on-Chip*,
DOI 10.1007/978-1-4419-6911-8_10, © Springer Science+Business Media, LLC 2011

power consumption and communication latencies observed among large numbers of cores will vastly impact the overall system performance. One commonly suggested communication scheme is to connect them through the NoC and send data using package switching [2, 3]. Recent NoC design efforts include Intel's 80-core design [4] on a single chip and Tilera's 64-core microprocessor [5], where processing cores are homogenous in both designs.

The future trend for NoCs, however, will be heterogeneous in nature. We expect that some cores will be general purpose processors running at moderate clock rates with normal supply voltages for achieving higher data processing rates, while others will be application-specific processors running at near/sub-V_{th} modes with much lower clock rates and lower supply voltages. For such heterogeneous many-core systems, the on-chip interconnect network has been projected as the primary performance bottleneck [6–9] to the nanometer processor in terms of power and latency. We advocate the use of reconfigurable interconnect as a means of providing power-efficient adaptation of the interconnect among various components in a heterogeneous many-core design.

In particular, we propose the use of low power multiband RF-interconnect (RF-I) that can concurrently communicate via multiple frequency bands using shared transmission lines to provide effective speed-of-light signal transmission, low power operation, and reconfigurable bandwidth. Effectively, RF-I provides a flexible set of low-latency communication channels that can be adaptively configured to the bandwidth demands of a particular architecture – providing a number of concurrent virtual communication channels out of a shared physical transmission media, such as on-chip transmission lines. We also investigate the CMOS mixed-signal circuit design challenges to bring RF-I to fruition and offer physical design examples to ensure RF-I's signaling integrity and efficiency. Furthermore, we propose a micro-architectural exploration framework that can be used to facilitate the exploration of scalable architectures based on physical planning and prototyping, particularly for a large number of processing cores. Our previous work has considered an architecture that combines a mesh topology implemented with conventional interconnect that is overlaid with a RF-I transmission line bundle. The RF-I acts like a reconfigurable superhighway, providing flexible, accelerated communication channels for critical/sensitive communications. The conventional interconnect acts as a more general set of surface streets that extend communications to all components on a chip.

10.2 Interconnect Problem in Future Information Processor

The contemporary solution to building many-core on-chip interconnects is the use of CMOS repeaters. However, despite improvements in transistor speed from one technology generation to the next, wire resistance and capacitance scale poorly [9, 10], if at all. Figures 10.1 and 10.2 project the performance of a 2-cm on-chip repeater buffer link (i.e., a modern 2×2 cm^2 CMP inter-core interconnect) from 130 nm to 16 nm CMOS technology. These figures demonstrate that the link delay

Fig. 10.1 Non-scalable delay of RC repeater buffer

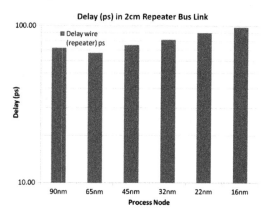

Fig. 10.2 Slow energy per bit scaling of RC repeater buffer

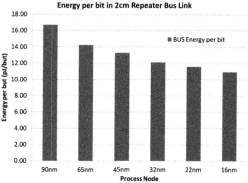

will grow worse with shrinking feature sizes, and the scaling of energy per bit will be saturated at about 10 pJ/bit. One possible solution is to use low-voltage swing interconnects [11–13], which inevitably require a power-hungry equalizer due to the severe dispersive channel characteristics of the on-chip wire across base-band frequencies. The signal bandwidth of the existing RC repeater buffer operates at not more than 5 GHz in the foreseeable future, which is primarily due to severe thermal and power constraints.

As shown in Fig. 10.3, an RC repeater buffer only utilizes less than 2% of the maximum available bandwidth, set by the cutoff frequency f_T of CMOS, which is 240 GHz in 45 nm CMOS today and will eventually reach 600 GHz in 16 nm CMOS according to the ITRS [14]. Owens et al. [7] even predicted that at 22 nm technology, the total network power using repeater buffers will dominate chip-multiprocessor (CMP) power consumption. Consequently, future CMPs using the RC repeater buffer would encounter serious communication congestion and spend most of their time and energy in "talking" instead of "computing". Intel's 80-tile CMP [4] demonstrated that their NoC consumed 30% of the total 100 W power consumption for a 10×8 mesh NoC running at a 4 GHz clock to support the 256 GB bisection bandwidth that is crucial for massive parallel processing. The same CMP design also requires 75 clock cycles in the worst case for a data packet to communicate between two opposite corners of the die. This clearly reveals the need to develop new on-chip

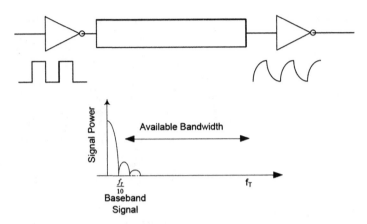

Fig. 10.3 Given data rate of 4 Gbit/s and f_T at 240 GHz in 45 nm CMOS, the RC repeater buffer only utilize 2% of maximum available bandwidth

interconnect schemes that are both scalable in energy consumption and efficient in inter-core communication.

10.3 How Can RF Help?

Interconnect fabric for future computing systems must not only be capable to have high performance with given low power budget but also adaptive according to individual processing core needs. As we have pointed out, the traditional repeater buffer does not fulfill such requirements due to its poor performance scaling in general and poor noise immunity. It is also not reconfigurable to perform multicast for network communications without a large overhead and cannot be adapted dynamically to allocate the changing needs of bandwidth. To circumvent the above deficiencies in traditional baseband-only type of interconnect, we propose to use the multiband RF-interconnect for reasons detailed as follows.

One of the key benefits of the scaling of CMOS is that the switching speed of the transistor improves over each technology generation. According to ITRS [14], f_T and f_{max}, will be 600 GHz and 1 THz, respectively, in 16 nm CMOS technology. A new record of a 324 GHz millimeter-wave CMOS oscillator [15] has also been demonstrated in standard digital 90 nm CMOS process. With the advance in CMOS mm-wave circuits, hundreds of gigahertz bandwidth will be available in the near future. In addition, compared with CMOS repeaters charging and discharging the wire, EM waves travel in a guided medium at the speed of light which is about 10 ps/mm on silicon substrate. The question here is: how can we use over hundreds of GHz of bandwidth in a future mobile system through RF-I while concurrently achieving ultra-low power operation and dynamic allocation in bandwidth to meet future heterogeneously integrated mobile system needs?

One of the possibilities is to use multiband RF-I, based on frequency-division-multiple-access algorithms (FDMA) [16–20, 22, 23], to facilitate inter-core communications on-chip. In the past, we have already demonstrated such interconnect schemes both on-chip and 3DIC (i.e., three dimensional integrated circuit) that RF-interconnects can achieve high speed (5–10 Gb/s in 0.18μm CMOS), low BER (10^{-14} without error correction) [17, 18], seamless re-configurability, and simultaneous, communications between multiple I/O users via multiple frequency bands using shared physical transmission lines. The main advantages of RF-I include:

- Superior signal to noise ratio: Since all data streams modulate RF-carriers, which are at least 10 GHz above the baseband, the high speed RF-interconnect does not generate and/or suffer from any baseband switching noise. This reduces possible interference to the sensitive near/sub-V_{th} operated circuit.
- High bandwidth: A multiband RF-interconnect link has a much higher aggregate data rate than a single repeater buffer link.
- Low power: Compared to a repeater buffer, a multiband RF-interconnect is able to operate at much better energy per bit in the NoC. Compared to normal repeated wire networks, which consume considerable amounts of power, a few RF-I nodes only consume a very small amount of power (see Sect. 10.4, benchmarked using pJ/bit as a metric).
- Low overhead: High data rate/wire and low area/Gigabit and low latency due to speed-of-light data transmission (see Sect. 10.4, benchmarked using Area/(Gbit/sec) as a metric).
- Re-configurability: Efficient simultaneous communications with adaptive bandwidths via shared on-chip transmission lines.
- Multicast support: Scalable means to communicate from one transmitter to a number of receivers on chip.
- Total compatibility and scalability: RF-I is implemented in mainstream digital CMOS technology which can directly benefit from scaling of CMOS.

The concept of RF-I is based on transmission of waves, rather than voltage signaling. When using voltage signaling in conventional RC time constant dominated interconnects, the entire length of the wire has to be charged and discharged to signify either '1' or '0'. In the RF approach, an electromagnetic (EM) wave is continuously sent along the wire (treated as a transmission line). Data are modulated onto that carrier wave using amplitude and/or phase changes. One of the simple modulation schemes for this application is binary-phase-shift-keying (BPSK) where the binary data changes the phase of the wave between $0°$ and $180°$. By expanding the idea of the single carrier RF-I, it is possible to improve bandwidth efficiency using N-channel multi-carrier RF-I. In multi-carrier RF-I, there are N mixers in the Tx. Each mixer up-converts individual base-band data streams into a specific channel. Those N distinct channels transmit N different data streams onto the same transmission line. The total aggregate data rate (R_{Total}) equals to $R_{Total} = R_{baseband} \times N$, where the data rate of each base-band is $R_{baseband}$ and

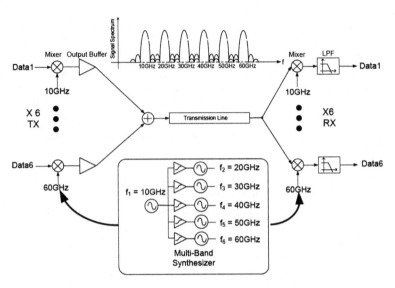

Fig. 10.4 Conceptual schematic of the multi-band RF-Interconnect

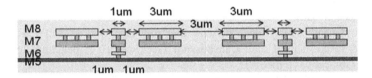

Fig. 10.5 Exemplary cross-section of the on-chip differential transmission line

the number of channels is N. A conceptual illustration of a six-carrier FDMA RF-interconnect is shown in Fig. 10.4.

Future on-chip RF-I will require on-chip transmission-lines (TLs) that can achieve multiband communication with high aggregate data rates, low latency, low signal loss, low dispersion, and compact Si-area. One particular challenge is to simultaneously support both baseband and RF bands on a single TL without severe inter-channel interference. In this case, two fundamental propagation modes of wave in the TL, odd and even modes, are used to support base band and RF-band, respectively. Since the odd mode and even mode are orthogonal, we design a new type of on-chip transmission line that can support dual-mode wave propagation. The new design combines both differential and coplanar transmission line structure–the cross-section of the TL is illustrated in Fig. 10.5. The top two thick metal layers (M7 and M8) act as a differential signal line to support high frequency RF-band data in the odd mode, while the M5 layer acts as ground plan to support baseband data in the even mode. Verified using EM simulation, a simple side wall between two signal lines can reduce cross-coupling by 10 dB. The latency of such TLs is about 70 ps/cm and the loss is 15 dB/cm, both at 60 GHz.

10.4   Expected Performance of RF-I with Scaling

Future CMPs require scalable interconnects to satisfy future needs in communication bandwidth, power budget, and Si-area. In RF-I, the size of passive devices, such as inductors, is the dominant consumer of silicon area. Since the size of a passive device is inversely proportional to the operational frequency, as higher carrier frequencies are used , the size of the passive device can be greatly reduced. At 20 GHz, the size of the inductor is approximately $50\,\mu m \times 50\,\mu m$. However, due to wavelength scaling, the size of the inductor at 400 GHz can be as small as $12\,\mu m \times 12\,\mu m$, roughly a 20 × reduction in area. As long as the carrier frequency can increase at each new generation of technology, the transceiver area will also scale down. Switching as fast as 300 GHz (i.e., half of the f_T of 16 nm CMOS [14] to deliver reasonable gain) in future generations of CMOS will allow us to implement a large number of high frequency channels for a physical RF-I bus. In each new technology generation, the number of channels available on a single TL can be expected to grow. Nonetheless, the average power consumption per communication band is expected to stay constant (about 4–5 mW as seen in Table 10.1). The logic behind this assumption is that although RF circuits at higher carrier frequencies require more power, this additional power is compensated by the power saved at lower frequency communication bands due to higher operation frequency transistors available with scaling. In addition to more frequency bands, the modulation speed of each frequency carrier will also increase, allowing a higher data rate per band. As a result, the aggregate data rate is expected to increase by about 40% through every CMOS technology generation, as shown in Table 10.1. In addition, the cost of the data rate, in terms of area/(Gb/sec) and the energy consumption per transmitted bit are expected to scale down as well Figs. 10.6 and 10.7.

10.5   Implementation Examples

10.5.1   On-Chip Multi-Carrier Generation

For the on-chip RF-I illustrated in Fig. 10.4, a multiband synthesizer enables transmitting multiple bands of modulated RF signal on transmission lines using FDMA

Table 10.1   Scaling trend of RF-I

Technology	No. of Bands	Data rate per band (Gb/s)	Data rate per wire (Gb/s)	Energy per bit (pJ)	Area per Gbit (μm^2/Gbit)
90 nm	3RF + 1 BB	5	20	1.00	1640
65 nm	4RF + 1 BB	6	30	0.83	1183
45 nm	5RF + 1 BB	7	42	0.71	810
32 nm	6RF + 1 BB	8	56	0.63	562
22 nm	7RF + 1 BB	9	72	0.56	399
16 nm	8RF +1 BB	10	90	0.50	325

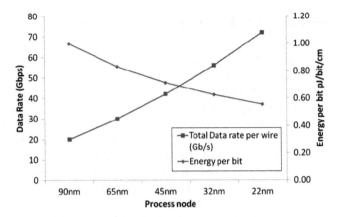

Fig. 10.6 RF-Interconnect scaling in terms of total energy per bit and total data rate per wire

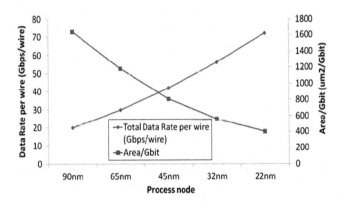

Fig. 10.7 RF-Interconnect scaling in terms of total data rate per wire and area per Gbps

between the transmitting and receiving units. A wide range on-chip frequency synthesis approach is thus required to enable the simultaneous generation of multiple carrier frequencies in the mm-wave range for multiband communications. Traditional approaches to on-chip frequency generation require dedicated VCOs and PLLs to cover multiple bands, thus consuming significant power and area. A new technique for generating multiple mm-wave carrier frequencies is proposed in our previous work [24] using simultaneous sub-harmonic injection locking to a single reference frequency. This concept is illustrated in Fig. 10.8. A master VCO generates a reference carrier at 10 GHz, which is fed into a differential pair. The differential pair generates the odd harmonic of the reference signal from the nonlinearity. The third harmonic of the reference carrier, 30 GHz, is then injected into the respective slave VCOs for them to lock on to the harmonic. The main advantages of this technique are reduction in power consumption, reduction in silicon area, and simpler carrier distribution networks. A prototype of 30 and 50 GHz sub-harmonic injection-locked VCOs was realized in a 90 nm digital CMOS process, as shown

Fig. 10.8 Schematic of the sub-harmonic injection locked VCO

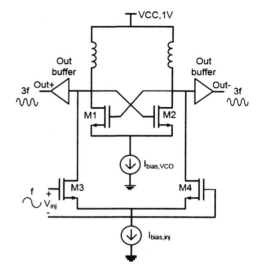

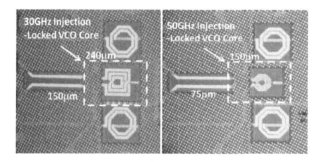

Fig. 10.9 Die photograph of the 30 GHz and 50 GHz sub-harmonic injection VCO

in Fig. 10.9, and able to lock on from the second to eighth harmonics of the reference frequency with locking range reaching 5.6 GHz. Simultaneous locking on to the third and fifth harmonics of a 10 GHz reference signal was also demonstrated, as shown in Fig. 10.10.

10.5.2 On-Chip RF-Interconnect

In this section, we illustrate the implementation of a simultaneous tri-band on-chip RF-interconnect [25] to demonstrate the feasibility of multiband RF-interconnect for future network-on-chip. In this design, two RF bands in mm-wave frequencies, 30 and 50 GHz, are modulated using amplitude-shift keying, while the base-band uses a low swing capacitive coupling technique. Each RF-band and base-band

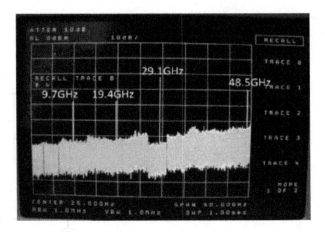

Fig. 10.10 Output spectrum of the 30 GHz and 50 GHz VCO simultaneously locked with the same reference source at 9.7 GHz

carries 4 and 2 Gb/s, respectively. Three different bands, up to 10Gb/s in total, are transmitted simultaneously across a shared 5 mm on-chip differential transmission line.

Like many other communication systems, signal to noise ratio of RF-I must be first estimated before starting on any major system designs such as selecting modulation scheme and designing transceiver architecture. From the SNR, we can estimate the bit error rate of the overall system. There are three types of noise that we should consider in RF-I. The first type of noise source is thermal noise from passive/active device. The second type of noise source is power supply noise. The third type of noise source is inter-channel interference.

Thermal noise from passive and active devices is one of the major sources of noise, which has been optimized to be low noise receiver front-end in many communication systems. RF-I, on the other hand, is not limited by thermal noise. We can simply deduce it from the following simple calculation.

We assume the transmitter using amplitude-shift-keying modulation with carrier at 60 GHz has 10% of output efficiency, and the total power consumption is 3 mW. The average output power, P_{TX}, will be

$$P_{TX} = 3 \, mW \times 10\% \times 0.5 = -8.24 \, dBm. \tag{10.1}$$

Based on full-EM wave simulation (measurement) on transmission line, the average signal attenuation of the on-chip transmission line is 1.5 dB/mm at 60 GHz. Assuming the average length of an on-chip transmission is 1 cm. The total signal loss will be −15 dB. Therefore, the signal power at the receiver front-end will be

$$P_{RX} = -8.2 \, dBm - 15 \, dBm = -23.2 \, dBm. \tag{10.2}$$

Assuming the channel bandwidth is 20 GHz and the noise figure of the receiver, NRX, is 10 dB. With that information, we can calculate the noise power at the receiver in the following:

$$P_{\text{noise}} = 4\,\text{KTR BW} + \text{NF} = -174\,\text{dB} + 10\log(\text{BW}) + \text{NF} = -61\,\text{dBm}. \quad (10.3)$$

After getting the signal power and noise power at the receiver front-end, we are ready to calculate the signal to noise ratio (SNR)

$$\text{SNR}_{\text{RX}} = P_{\text{signal}}(\text{dBm}) - P_{\text{noise}}(\text{dBm}) = 37\,\text{dB}. \quad (10.4)$$

Since the SNR at the receiver front end is 37 dB, the bit error rate of the on-chip RF-I is not limited by the thermal noise.

In future CMP, there will be over tens or even hundreds of processing cores in a single die. These noisy digital circuit generates noisy switching noise and couples to sensitive mixed-signal circuit through the power supply network and low impedance CMOS substrate. Therefore, rejecting digital switching noise becomes one of the most important design considerations. Fortunately, in RF-I, all data streams modulate RF-carriers, which has at least 10 GHz above the baseband, and thus the high speed RF-interconnect does not generate and/or suffer from any baseband switching noise which is usually below 10 GHz. Comparing to conventional on-chip interconnects technique, low-swing signaling, which is directly suffered from the digital supply noise, RF-I clearly has superior power supply noise rejection.

One of the advantages of RF-I is that it provides much higher aggregate data rate than conventional on-chip interconnect by sending multichannels of data simultaneously into one single transmission line. Interference among multiple channels become critical in RF-I design. One particular parameter to quantify channel interference is signal to interference ratio (SIR). Assuming the minimum SIR is 20 dB and the modulation scheme is amplitude-shift-keying (ASK). The power spectrum of the ASK is

$$P(f) = (A/2)^2\, T\, \text{sinc}^2\left(\frac{f - f_c}{T}\right) + (A/2)^2\, \delta(f - f_c). \quad (10.5)$$

From the power spectrum of the ASK, the separation between two adjacent channels must be at least $3\,\text{BW}_{\text{data}}$ to satisfy the 20 dB of SIR, where BW_{data} is the data rate of the data stream. For instance, the channel separation is 15 GHz for the data rate of 5Gbps in each channel.

The schematic of the proposed tri-band RF-I is shown in Fig. 10.11. The modulation scheme of each RF band is amplitude-shift keying (ASK), in which a pair of on–off switches directly modulates the RF carrier. Unlike other modulation schemes such as BPSK [18, 22, 23], the receiver of the ASK system only detects the changes in amplitude and not phase or frequency variations. Therefore, it operates asynchronously without a power hungry PLL. It also eliminates the need for coherent carrier regeneration at the receiver. Consequently, RF-I does not suffer from carrier

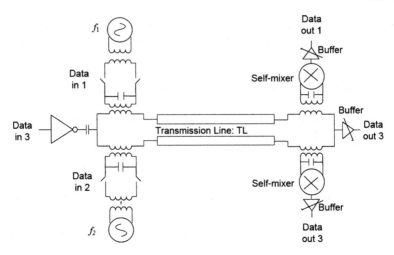

Fig. 10.11 Die photograph of the tri-band on-chip RF-I based

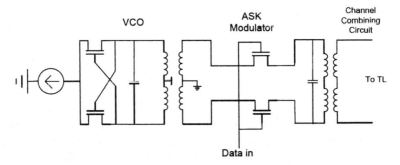

Fig. 10.12 Schematic of the transmitter of the RF band

variations between the transmitter and receiver due to process variation. Moreover, RF-I can also operate properly with conventional digital logic circuits placed directly under its passive structure, which gives better area utilization.

For each RF band, the design uses a minimal configuration that includes a voltage-controlled oscillator(VCO) and a pair of ASK switches on the transmitter side, as well as a self-mixer and baseband amplifiers on the receiver side. As shown in Fig. 10.12, the VCO generates the RF-carrier and acts as a push-pull amplifier. The RF-carrier from the VCO is first inductively coupled to the ASK modulator through a 2:1 ratio transformer. After that, the input data stream modulates the RF carrier via a pair of ASK switches. In order to maximize the modulation depth of the ASK signal, the size of switches is chosen to provide an optimal balance between the on-state loss and the off-state feed through. After the ASK modulation, the differential ASK signal is inductively coupled to the transmission line(TL) through the second frequency selective transformer. The impedance matching requirement is greatly relaxed because the reflected wave is attenuated significantly in the on-chip

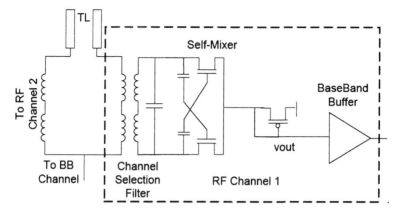

Fig. 10.13 Schematics of the RF receiver

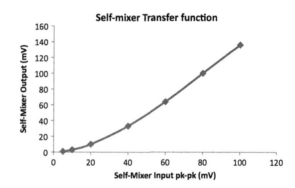

Fig. 10.14 Large Signal voltage transfer curve of the self-mixer

TL after reflection. By choosing the RF-carrier in mm-wave frequencies, the higher carrier to data rate ratio further minimizes the dispersion of the signal and removes the need for a power hungry equalization circuit. The receiver architecture in each RF band is shown in Fig. 10.13. The self-mixer acts as an envelope detector and demodulates the mm-wave ASK signal into a baseband signal, where it is further amplified to a full-swing digital signal. The simulated voltage transfer curve of the self-mixer is plotted in the Fig. 10.14 The ordinary measurement technique on frequency response in linear circuit, such as small signal AC response, is not applicable to the self-mixer, operating nonlinearly in nature. Figure 10.15 shows the simulated frequency response of the self-mixer by measuring the eye-opening at the output of self-mixer in different input data rate of the ASK signal. The simulated result shows that the self-mixer is able to demodulate the ASK signal as high as 10 Gbps. Figure 10.16 shows the transient simulation of the self-mixer running at 5Gbps ASK signal which has carrier at 60 GHz.

The baseband (BB) uses a low-swing interconnect technique using capacitive coupling [12]. As shown in Fig. 10.17, the baseband data is transmitted and received using the common mode of the differential TL. At low frequencies, the transformer

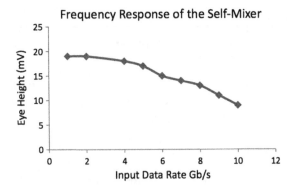

Fig. 10.15 Large Signal voltage transfer curve of the self-mixer

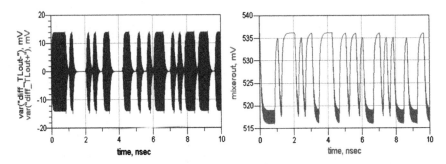

Fig. 10.16 Transient simulation of the self-mixer (*left*) input ASK modulated signal with carrier at 60 GHz and (*right*) the demodulated 10Gbps ASK signal

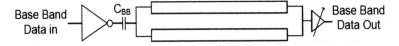

Fig. 10.17 Equivalent circuit of the base band in common mode

becomes a short circuit, and a pair of low-swing capacitive coupling buffers transmits and receives the baseband data at the center tap of the transformer.

The transmitter and the receiver are connected by an on-chip 5-mm long differential TL. In order to support simultaneous multiband RF-I on a shared TL, RF and BB are transmitted in differential mode and common mode, respectively. These two propagation modes are naturally orthogonal to each other and suppress the inter-channel interference (ICI) between RF and BB. Even with finite coupling between differential mode (RF) and the common mode (BB), the low-pass characteristic of the BB receiver and the band-pass characteristic of the RF receiver can provide further rejection of any possible ICI between RF and BB. The remaining challenge is the ICI between the different RF channels. In the transmitter, the frequency selectivity of the second transformer in each RF band reduces ICI due to

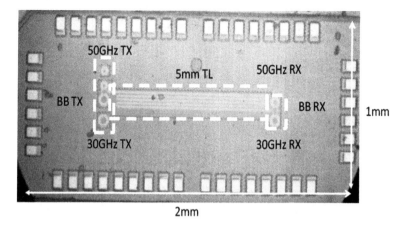

Fig. 10.18 Die photograph of the tri-band on-chip RF-I based

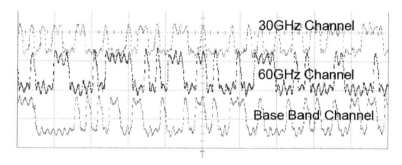

Fig. 10.19 Data output of the tri-band waveform 30 GHz, 50 GHz and base band

signal leakage to the adjacent RF band's ASK modulator. In the receiver of the RF bands, the transformer at the input of the self-mixer acts as a band-pass filter.

The tri-band on-chip RF-I is implemented in the IBM 90 nm digital CMOS process. The die size is 1 mm × 2 mm, as shown in Fig. 10.18. Figure 10.19 shows the recovered data waveform of the three bands: 30 GHz, 50 GHz and BB. The maximum data rates for each RF band and BB are 4 Gb/s and 2 Gb/s, respectively. The total aggregate data rate is 10 Gb/s. A RF-I with TX and TL only was also implemented for measuring the spectrum of the tri-band RF-I signals. A 67-GS Cascade Micro-Probe directly probes the on-chip differential TL (only differential mode can be measured). Figure 10.20a shows the free running VCO spectrum without input data modulation on both RF bands at 28.8 and 49.5 GHz respectively. When the two uncorrelated 4 Gb/s random data streams are applied to both RF bands, as shown in Fig. 10.20b, the spectrum of each band broadens and spreads over 10 GHz of bandwidth. The tri-band RF-I achieves superior aggregate data rate (10 Gb/s), latency (6ps/mm), and energy per bit 0.45 pJ bit and 0.625 pJ/bit/mm, for RF and BB, respectively, which is summarized in Table 10.2.

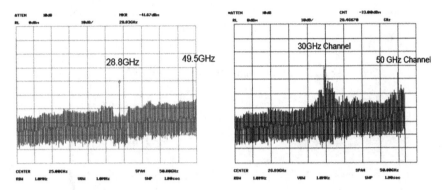

Fig. 10.20 Spectrum on the differential mode RF-I signal with (**a**) no data input (**b**) with 4 Gbps data input in each band

Table 10.2 Performance summary of the tri-band RF-I. *VCO power (5 mW) can be shared by all (many tens) parallel RF-I links in NOC and does not burden individual link significantly

	Tri-band RF-I
Interconnect technique	RF-I
Bands	30 GHz, 50 GHz, base band
Data rate in RF channel (Gbps)	4
Data rate in BB channel (Gbps)	2
Total aggregate data rate (Gbps)	10
BER	10^{-9} across all channels
Latency (ps/mm)	6
Energy per bit (RF) pJ/bit	0.45 (5mm)*
Energy per bit (BB) pJ/bit	0.63 (5mm)*

10.5.3 3D IC RF-Interconnect

One of the current technological trends in CMOS processes is three-dimensional stacking [9, 26], in which several thin tiers of circuitry are stacked vertically to achieve a higher level of integration. Due to vertical integration, the same functionality can be implemented in a smaller chip area, reducing both cost and the distance signals that are required to travel across the chip. Reduced distance decreases both transmission latency and the consumed energy. However, 3D stacking requires vertical connection between transistor and metal tiers, usually implemented using metal studs that cut through layers of silicon and insulators. Alignment of such direct connection is difficult on a large scale and therefore requires a relatively large connection area.

The use of RF signaling has an advantage over standard voltage signaling for inter-layer communication. Because the signal is modulated on a high-frequency carrier, it does not require a direct connection, and capacitive or inductive coupling is enough for transmission. Figure 10.21 shows a schematic view of a fabricated 3D

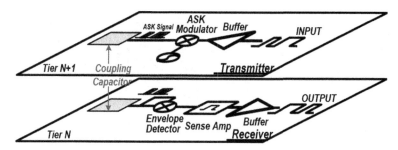

Fig. 10.21   Schematic of the RF-interconnect implemented in a 3D 0.18 μm CMOS process

Fig. 10.22 Chip photograph
of 3D RF-interconnect

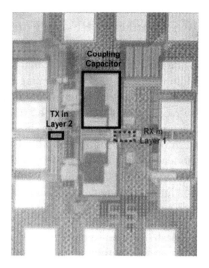

integrated circuit demonstrating an RF-interconnect using capacitive coupling, with the photograph of the actual die shown in Fig. 10.22. In this circuit implemented in 180 nm 3D SOI processed provided by the MIT Lincoln Lab [17], an amplitude shift keying (ASK) modulation of a 25 GHz carrier is used, so that recovery of the data requires only an envelope detector. Metal layers in each of the tiers are used to form capacitors with values of tens of femto-farads that are sufficient for effective coupling. This realized RF-interconnect achieves a maximum data rate of 11 Gb/s per wire and a very low bit error rate (BER) of 10^{-14} measured at about 8 Gb/s, as shown in Fig. 10.23. Based on estimation, separation distance between adjacent channel can be as small as 6.5 times of the separation distance between layer for 10^{-12} BER. For example, in 180 nm 3D SOI process, the separation distance between two layer is 3 μm and the separation distance between adjacent channel in capacitive coupling interconnect is about 20 μm which is only about two-third of the inductive coupling interconnect [21]. Therefore, the use of small capacitors for coupling has an advantage over on-chip inductors or antennas due to the better field confinement that reduces cross-talk and interference between differential links.

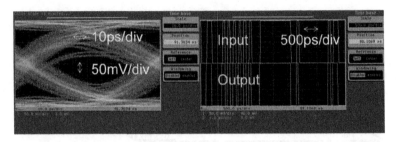

Fig. 10.23 Measurement result of the 3D RF-Interconnect at 11 Gbps with 2^{15}-1 PRBS

10.6 Impact of RF-I in Future SoC/NoC Architecture

While RF-I has dramatic potential in terms of low-latency, low-power, high-band-width operation, the key enabling component of RF-I for future microprocessor architectural design is reconfigurability. As an example of this reconfigurability, we recently proposed MORFIC (mesh overlaid with RF inter-connect) [19, 27], a hybrid NoC design which is shown in Fig. 10.24. It is composed of a traditional mesh of routers augmented with a shared pool of RF-I that can be configured as shortcuts within the mesh. In this design, we have 64 computing cores, 32 cache memory modules, and four memory output ports – and RF-I is a bundle of transmission lines spanning the mesh and features 16 carrier frequencies. We examined four architectures:

1. Mesh baseline – a baseline mesh architecture without any RF-I;
2. Mesh wire baseline – the baseline mesh architecture with express shortcuts between routers (conventional wire, not RF-I) that are chosen at chip design time (i.e., no adaptability to application variation);
3. Mesh static shortcuts – the same express shortcuts as the Mesh Wire Baseline but using RF-I instead of conventional repeated wire;
4. Mesh adaptive shortcuts – the overlaid RF-I with shortcuts tailored to the particular application in execution.

From the simulation results of our in-house cycle-accurate simulator [28], we demonstrated a significant performance improvement of the mesh adaptive short-cuts over the mesh baseline, an average packet latency reduction of 20–25% [19], through the reconfigurable RF-I, as shown in Fig. 10.25. We further demonstrated a 65% power reduction [27] by reducing the bandwidth of the baseline mesh by 75% – reducing the 16 Byte wide to 4 Byte wide baseline mesh, as illustrated in Fig. 10.26. Our continued exploration of the MORFIC architecture will be instrumental in gauging future CMP interconnect design tradeoffs, and in better quantifying what benefits CMPs can expect from MORFIC in future generations of CMOS technologies down the road.

Fig. 10.24 Schematic of
MORFIC

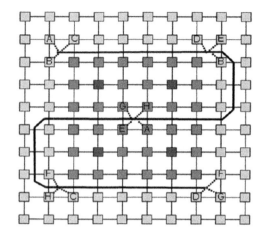

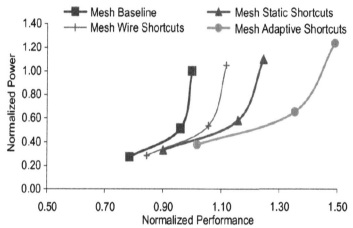

Fig. 10.25 Power performance trade-off curve of the mesh adaptive shortcuts over the mesh baseline

10.7   Future RF-I Research Direction

Before addressing possible future research directions for RF-I, we should compare the performance and the proper communication range for all three types of interconnects, including the traditional parallel repeated wire bus, the RF-I, and the optical interconnect. We first compare the latency, the energy consumption per bit, and the data rate density among them in Fig. 10.27 for the same 2-cm communication distance on-chip. The performance of the parallel repeater bus is projected according to the ITRS digital technology roadmap [14] with optimized repeater design practice [29]; the RF-I performance is estimated based on the RF-technology roadmap, employing our proposed RF-I design methodology portrayed above. The optical interconnect performance is calculated based on [30] and extrapolated to

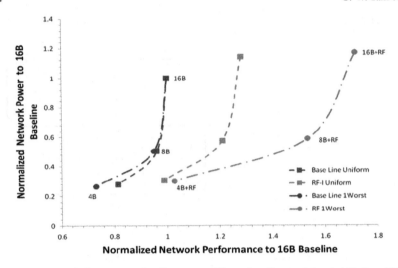

Fig. 10.26 Power performance trade-off curve on different baseline mesh bandwidth from 16 byte wide to 4 byte wide baseline mesh

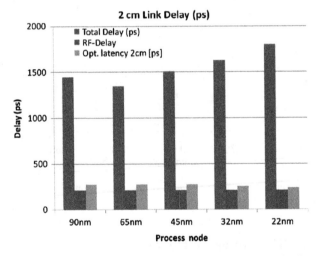

Fig. 10.27 Comparison of Interconnect technologies for a global 2 cm on-chip distance of latency between a traditional repeated parallel bus, RF-I and optical interconnect

further scaled technology nodes. In contrast to the latency increase of the traditional repeater bus against the scaling shown in Fig. 10.28, RF and optical interconnects are able to maintain similarly low latency over the scaling and keep the 2-cm data transmission within a clock cycle. The RF and optical interconnects again show significant benefit in energy consumption over the traditional bus, as shown in Fig. 10.29. The RF-I even scales slightly better than that of optical interconnect in terms of absolute energy per bit. Data rate density is expected to improve in all

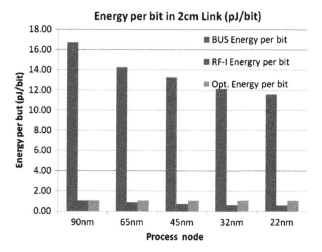

Fig. 10.28 Comparison of Interconnect technologies for a global 2cm on-chip distance of energy consumption per bit for a traditional repeated parallel bus, RF-I and optical interconnect

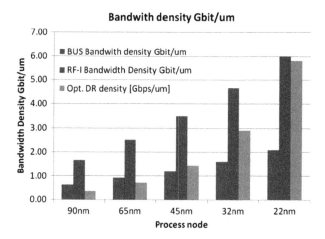

Fig. 10.29 Comparison of Interconnect technologies for a global 2cm on-chip distance of data rate density for a traditional repeated parallel bus, RF-I and optical interconnect

three interconnects: The bus would benefit from the wire pitch; RF-I benefits from the number of carrier bands and the effective transmission speed possible; and the optical data density should improve under the assumption of more wavelengths used [31], although its optical transceiver typically requires non-CMOS devices which are less-scalable due to fundamental physical constraints and often more sensitive to temperature variations. RF-I, on the other hand, has the major advantage of using the standard digital CMOS technology.

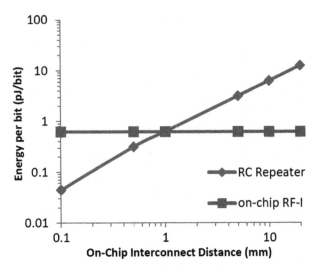

Fig. 10.30 RF-I will crossover the energy efficient curve of the RC repeater and become more energy efficient above a 1 mm interconnect distance at a 16 nm CMOS process

Besides the performance, we may also assess the optimized communication range for each of the interconnect technologies. As CMOS continues to scale toward 16 nm, traditional on-chip RC repeated wires are more suitable for local interconnects with short communication distance due to further increased physical density through the use of minimum-feature-width metal wires [10]. Figure 10.30 illustrates the projected power/performance of both RC wires with optimal delay [14, 29] and RF-I with a 16 nm CMOS process. Under approximately 1mm, the RC repeater is able to provide superior energy efficient communication, but beyond 1 mm, the repeater buffers become less efficient than those of RF-I. The RF-I is expected to maintain its performance advantages for global interconnect on-chip due to its total compatibility with the CMOS technology, but can it maintain the same superiority to an extended distance off-chip? Especially, to what range can it compete with the optical interconnect which is clearly superior for longer-distance communications? We offer the answer to those questions by comparing the energy efficiency between the off-chip RF-I and optical interconnect in Fig. 10.31, where the off-chip RF-I energy-per-bit is estimated with the physical transceiver/transmission line designs based on [18], and the optical interconnect results are obtained through the data from [32, 33]. Accordingly, the RF-I actually exhibits better energy efficiency at midrange distances of 30 cm or below. As the communication distance increases, RF-I energy efficiency decreases rapidly due to the excessive power required to compensate for the severe loss from the off-chip printed-circuit-board transmission lines, while the power consumption of optical interconnect remains almost constant. Therefore, despite substantial disadvantages in integration and cost, the optical interconnect becomes more beneficial at interconnect distances beyond 30 cm.

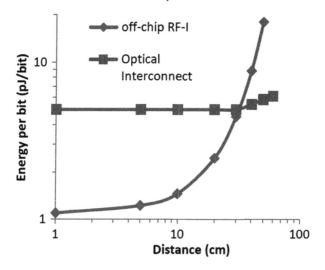

Fig. 10.31 RF-I has much better energy efficiency in the midrange distance of 30 cm or below, while the optical interconnect does not have any benefit until an interconnect distance over 30 cm

Fig. 10.32 Communication range versus interconnect technologies as CMOS process continuously to scale toward 16 nm

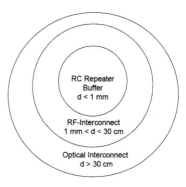

That is to say, in between traditional RC repeater buffer and optical interconnects, there is an obvious technology gap for achieving cost/performance-effective communications in mid-distance range from a few millimeters to several tens of centimeters. The CMOS compatible RF-I may be the right technology to fill in such a technology gap, as shown in Fig. 10.32, with the lowest latency, the least energy consumption, and the highest data rate density.

However, in order to take full advantage of its potential and to be adopted by the mainstream industry for CMP implementations, we must further advance RF-I circuitry and low power many-core architecture designs in the following areas:

- Effective channel allocation scheme to support co-existing RF-band and base band on a shared transmission line. The particular challenges include designing a multiband coupler which is small in area, high in coupling efficient, and yet minimizes the inter-channel interference.

- Reliable signaling techniques to provide an interference and noise resilient RF-I – in future RF-I for NoC, highly reliable interconnects are required such that the bit-error rate (BER) is sufficiently low to maintain reliable computing. The BER in the current tri-band RF-I design [25] may not meet the future requirements due to interference from noisy digital circuits and the thermal noise from the active device.
- Transceiver architecture that can support self-arbitrated and collision-free multi-cast communications – one of the potential advantages of RF-I is to provide effective broadcasting over the NoC. However, protocol and infrastructure supporting effective self-arbitration and collision-free multi-casting are not well developed yet.
- Adaptive loading balancing of the NoC through RF-I – our current designs re-configure the NoC at a coarse granularity, leveraging phase locality within the application to amortize the cost of reconfiguration over many cycles. However, further gain may be possible with more dynamic adaptation – such a design will require a mechanism to rapidly arbitrate RF-I frequencies among multiple communicating components and rapidly notify these components of their communicating frequencies.
- Reduction of NoC's memory bandwidth, latency, and power limitations by fully using RF-I in the memory hierarchy – while we have considered overlaid RF-I for express channels in a mesh topology, there are further potential gains that can be realized using RF-I as the main communication channel in the NoC. For example, we are exploring RF-I-enabled cross-bars and RF-I-based cache partitioning.
- Leveraging multi-cast to improve cache coherence, transactional memory, thread-level synchronization, or composable cores – our initial efforts have dealt with coarse-grain arbitration for multi-cast masters among a small set of potential senders, but we are also considering a larger scale implementation which can enable more nodes to cooperate as multi-cast senders. Such an implementation can dramatically improve the performance of more sophisticated cache coherence protocols that require collective communication, transactional memory schemes that require commits to be broadcast to all participating cores, synchronization techniques such as barriers in multithreaded applications, and composable cores where a number of simple cores cooperate together to handle a single sequential thread. In this latter case, RF-I has a dramatic potential to accelerate communication between cooperating cores.
- Transmission line base RF-I is difficult to scale more than 1,000-core NoC – In the case of over 1,000-core NoC, transmission line needs to span the entire chip area and requires excessive branching points to connect to local cores. One possible solution is on-chip wireless interconnect, in which frequency band up to sub-terahertz (100 GHz to 500 GHz). Lee [34] proposed a micro wireless interconnect architecture for NoCs with hundreds to thousands of cores which uses a two-tiered hybrid structure, wireless backbone and wired edges, to interconnect thousands of cores in NoCs. This new micro on-chip wireless interconnect eliminates long wires and reduces latency for long-haul, many-hop, inter-core communication. Moreover, based on simulation results, the latency of such two-tiered hybrid structure is reduced about 20–45%.

Acknowledgements The authors would like to thank the US DARPA and GSRC for their contract supports and TAPO/IBM for their foundry service.

References

1. S. Borkar, "Thousand Core Chips – A Technology Perspective," Proceeding of the 44th annual conference on design automation, pp. 746–749, 2007
2. W.J. Dally, B. Towles, "Route Packets, Not wire: On-Chip Inter-connection Networks," Proceeding of the 38th Design Automation Conference (DAC), pp. 684–689, 2001
3. L. Benini, G. De Micheli, "Networks on Chips: a new SoC paradigm," IEEE Computer Magazine, pp. 70–78, Jan. 2002
4. S. Vangal et al., "An 80-Title 1.28 TFLOPS Network-on-Chip in 65nm CMOS," IEEE International Solid-State Circuits Conference (ISSCC) Digest of Technical Papers, pp. 98–99, 2007, San Francisco, California, USA
5. S. Bell et al., "TILE64TM Processor: A 64-Core SoC with Mesh Intercon-nect," IEEE International Solid-State Circuits Conference (ISSCC) Digest of Technical Papers, pp. 88–89 2008, San Francisco, California, USA
6. R. Kumar, V. Zyuban, D. Tullsen, "Interconnections in multi-core architectures: Understanding Mechanisms, Overheads and Scaling," Proceed-ing of the 32nd International Symposium on Computer Architecture, pp. 408–419, June 2005
7. J.D. Owens, W.J. Dally, R. Ho, D.N. Jayasimha, S.W. Keckler, L.-S. Peh, "Research Challenges for On-Chip Interconnection Networks," IEEE MICRO, pp. 96–108, Sept 2007
8. T. Karnik, S. Borkar, "Sub-90nm Technologies-Challenges and Opportunities for CAD," Proceedings of International Conference on Com-puter Aided Design, pp. 203–206, November 2002
9. J. Cong, "An Interconnect-Centric Design Flow for Nanometer Technologies," Proc. of the IEEE, April 2001, vol. 89, no. 4, pp. 505–528
10. R. Ho, K.W. Mai, M. Horowitz, "The future of wires," Proceedings of the IEEE, vol. 89, no. 4, pp. 490–504, April 2001
11. A.P. Jose, K.L. Shapard, "Distributed Loss-Compensation Technique for Energy-Efficient Low-Latency On-Chip Communication," IEEE Journal of Solid State Circuits, vol. 42, no. 6, pp. 1415–1424, 2007
12. R. Ho et al., "High Speed and Low Energy Capacitively Driven On-Chip Wires," IEEE Journal of Solid State Circuits, vol. 43, no. 1, pp. 52–60, Jan 2008
13. H. Ito et al., "A 8-Gbps Low Latency Multi-Drop On-Chip Transmission Line Interconnect with 1.2mW Two-Way Transceivers," Proceeding of the VLSI Symposium, pp. 136–137, 2007
14. "International Technology Roadmap for Semiconductors," Semiconductor Industry Association, 2006
15. D. Huang et al., "Terahertz CMOS Frequency Generator Using Linear Superposition Technique," IEEE Journal of Solid State Circuits, vol. 43, no.12, pp. 2730–2738, Dec 2008
16. M.-C.F. Chang et al., "Advanced RF/Baseband Interconnect Schemes for Inter- and Intra-ULSI communications," IEEE Transactions on Electron Devices, vol. 52, no. 7, pp. 1271–1285, July 2005
17. Q. Gu, Z. Xu, J. Ko, M.-C.F. Chang, "Two 10Gb/s/pin Low-Power Interconnect Methods for 3D ICs," Solid-State Circuits Confe-rence, 2007. ISSCC 2007. Digest of Technical Papers. IEEE International, pp. 448–614, 11–15 Feb. 2007
18. J. Ko, J. Kim, Z. Xu, Q. Gu, C. Chien, M.F. Chang, "An RF/baseband FDMA-interconnect transceiver for reconfigurable multiple access chip-to-chip communication," Solid-State Circuits Conference, 2005. Digest of Technical Papers. ISSCC. 2005 IEEE International, pp. 338–602 vol. 1, 10–10 Feb. 2005
19. M.F. Chang, J. Cong, A. Kaplan, M. Naik, G. Reinman, E. Socher, S.-W. Tam, "CMP Network-on-Chip Overlaid With Multi-Band RF-Interconnect," IEEE International Conference on High Performance Computer Architecture Sym, pp. 191–202, Feb. 2008

20. M.F. Chang et al., "RF/Wireless Interconnect for Inter- and Intra-chip Communication," Proceedings of the IEEE, vol. 89, no. 4, pp. 456–466, April 2001
21. N. Miura, D. Mizoguchi, M. Inoue, K. Niitsu, Y. Nakagawa, M. Tago, M. Fukaishi, T. Sakurai, T. Kuroda, "A 1 Tb/s 3 W Inductive-Coupling Transceiver for 3D-Stacked Inter-Chip Clock and Data Link," IEEE Journal of Solid-State Circuits, vol. 42, no. 1, pp. 111–122, Jan. 2007
22. R.T. Chang, N. Talwalkar, C.P. Yue, S.S. Wong, "Near speed-of-light signaling over on-chip electrical interconnects," IEEE Journal of Solid-State Circuits, vol. 38, no. 5, pp. 834–838, May 2003
23. B.A. Floyd, C.-M. Hung, K.K. O, "Intra-chip wireless interconnect for clock distribution implemented with integrated antennas, receivers, and transmitters," IEEE Journal of Solid-State Circuits, vol. 37, no. 5, pp. 543–552, May 2002
24. S.-W. Tam et al., "Simultaneous Sub-harmonic Injection-Locked mm-Wave Frequency Generators for Multi-band Communications in CMOS," IEEE Radio Frequency Integrated Circuits Symposium, pp. 131–134, 2008
25. S.-W. Tam et al., "A Simultaneous Tri-band On-Chip RF-Interconnect for Future Network-on-Chip," VLSI Circuits, 2009 Symposium on, vol., no., pp. 90–91, 16–18, June 2009
26. J.A. Burns et al., "A Wafer-Scale 3-D Circuit Integration Technology," IEEE Transactions on Electron Devices, vol. 53, no. 10, pp. 2507–2516, October 2006
27. M.F. Chang et al., "Power Reduction of CMP Communication Networks via RF-Interconnects," Proceedings of the 41st Annual International Symposium on Microarchitecture (MICRO), Lake Como, Italy, pp. 376–387, November 2008
28. J. Cong et al., "MC-Sim: An Efficient Simulation Tool for MPSoC Designs," IEEE/ACM International Conference on Computer-Aided Design, pp. 364–371, 2008
29. J. Rabaey, A. Chandrakasan, B. Nikolic, "Digital Integrated Circuits: A Design Perspective," 2/e, Prentice Hall, 2003
30. N. Kirman et al., "Leveraging Optical Technology in Future Bus-based Chip Multiprocessors," 39th International Symposium on Microarchitecture, pp. 495–503 December 2006
31. M. Haurylau, G. Chen, H. Chen, J. Zhang, N.A. Nelson, D.H. Al-bonesi, E.G Friedman, P.M. Fauchet, "On-Chip Optical Interconnect Road-map: Challenges and Critical Directions," IEEE Journal of Selected Topics in Quantum Elec-tronics, vol. 12, no. 6, pp. 1699–1705, Nov. – Dec. 2006
32. H. Cho, P. Kapur, K. Saraswat, "Power comparison between high-speed electrical and optical interconnects for interchip communication," Journal of Lightwave Technology, vol. 22, no. 9, pp. 2021–2033, Sep. 2004
33. L. Schares, et.al., "Terabus: Terabit/Second-Class Card-Level Optical Inter-connect Technologies," IEEE Jour-nal of Selected Topics in Quantum Electronics, vol. 12, no. 5, pp. 1032–1044, Sept. – Oct. 2006
34. S.-B. Lee, et.al., "A Scalable Micro Wireless Interconnect Structure for CMPs," ACM MOBICOM 2009, pp. 217–228, 20–25 September 2009

Index